Stochastic Adaptive Search for Global Optimization

Nonconvex Optimization and Its Applications

Volume 72

STOCHASTIC ADAPTIVE SEARCH FOR GLOBAL OPTIMIZATION

ZELDA B. ZABINSKY
University of Washington
Seattle, Washington, USA

Springer Science+Business Media, LLC

Electronic Services <http://www.wkap.nl>

Library of Congress Cataloging-in-Publication

Zabinsky, Zelda B.
Stochastic Adaptive Search for Global Optimization
ISBN 978-1-4613-4826-9 ISBN 978-1-4419-9182-9 (eBook)
DOI 10.1007/978-1-4419-9182-9

Printed on acid-free paper.

I dedicate this book to
my parents, Joe and
Helen Zabinsky, to
show my love and
appreciation.

Contents

List of Figures

List of Tables

Preface

The field of global optimization has been developing at a rapid pace. There is a journal devoted to the topic, as well as many publications and notable books discussing various aspects of global optimization. This book is intended to complement these other publications with a focus on stochastic methods for global optimization.

Stochastic methods, such as simulated annealing and genetic algorithms, are gaining in popularity among practitioners and engineers because they are relatively easy to program on a computer and may be applied to a broad class of global optimization problems. However, the theoretical performance of these stochastic methods is not well understood. In this book, an attempt is made to describe the theoretical properties of several stochastic adaptive search methods. Such a theoretical understanding may allow us to better predict algorithm performance and ultimately design new and improved algorithms.

This book consolidates a collection of papers on the analysis and development of stochastic adaptive search. The first chapter introduces random search algorithms. Chapters 2-5 describe the theoretical analysis of a progression of algorithms. A main result is that the expected number of iterations for pure adaptive search is linear in dimension for a class of Lipschitz global optimization problems. Chapter 6 discusses algorithms, based on the Hit-and-Run sampling method, that have been developed to approximate the ideal performance of pure random search. The final chapter discusses several applications in engineering that use stochastic adaptive search methods.

The target audience includes graduate students, researchers and practitioners in operations research, engineering, and mathematics. A background in mathematics is assumed, as well as a knowledge of probabilistic concepts, such as Markov chains and moment generating functions.

It is possible for readers to skip the proofs and technical details and still grasp the basic ideas.

I apologize ahead of time for errors and inconsistencies in this book. I would appreciate help in correcting mistakes, so please email your suggestions to me at <zelda@u.washington.edu>. I will maintain a web page of errata at http://faculty.washington.edu/zelda/.

I would like to express my thanks to Professor Robert L. Smith for his teaching, mentoring, and continued collaboration. Many of the ideas in this book originated with him. I am also grateful to Professor Graham R. Wood for his inspiration, collaboration and insightful comments. I appreciate the encouragement I received from Dr.s Reiner Horst and Panos Pardalos to undertake this project, and Mr. John Martindale's patience. Dr. Birna Kristinsdottir has been extremely helpful throughout the stages of writing this book. Dr. Mirjam Dür and Dr. David Bulger have given valuable advice during the final stages of the writing. I thank Mr. Kyle Knopp for assisting me with the figures. Many other colleagues have been extremely helpful, and I particularly want to thank (in alphabetical order): Dr.s Bill Baritompa, Vladimir Brayman, Bruce Campbell, Tibor Csendes, Charoenchai Khompatraporn, Victor Korotkich, Wen Luo, Sudipto Neogi, János Pintér, Edwin Romeijn, Vesna Saviç, Yanfang Shen, and Mark Tuttle.

Support from the Industrial Engineering Program at the University of Washington is gratefully acknowledged, as well as a sabbatical in the Department of Mathematics and Computing at Central Queensland University (1996), an Erskine Fellowship from the Department of Mathematics at the University of Canterbury (1998), two National Science Foundation grants (DMI-9622433 for 1996-1999 and DMI-9820878 for 1999-2003), and participation in the Marsden Fund administered by the Royal Society of New Zealand (1999-2002).

I do not have enough words to express my thanks to my husband, Dr. John Palmer, and our children, Rebecca and Aaron, for their patience and enduring support. John has been a reader, prodder, and inspiration for this work. Without him, this book would not have been possible.

ZELDA B. ZABINSKY

SEATTLE, WASHINGTON, APRIL 2003

Chapter 1

INTRODUCTION

Global optimization refers to a mathematical program, which seeks a maximum or minimum objective function value over a set of feasible solutions. The adjective "global" indicates that the optimization problem may be very general in nature; the objective function may be nonconvex, nondifferentiable, and possibly discontinuous over a continuous or discrete domain. A global optimization problem with continuous variables may contain several local optima or stationary points. The problem of designing algorithms that obtain global solutions is very difficult when there is no overriding structure that indicates whether a local solution is indeed the global solution.

Even though global optimization problems are difficult to solve, applications of global optimization problems are prevalent in engineering and real world systems. Applications include engineering design in mechanical, civil, and chemical engineering, structural optimization, molecular biology and molecular architecture, VLSI chip design, image processing, and a number of combinatorial optimization problems [72, 185]. Engineering functions included in a global optimization problem are often supplied as "black box" functions, which might be a subroutine that returns a function evaluation for a specified solution. An engineering example of this type of problem is to minimize the weight of a structure, while limiting strain to be below a certain threshold [183]. Engineers must often provide some solution to their problem, even if it is a suboptimal one. Sometimes the global optimization problem may be so computationally difficult, that a practitioner is satisfied with any feasible solution. Optimization is currently being applied to complex systems where the objective function and constraints may only be evaluated using a simulation model. Not only does the problem lack structure and fall into the category of "black box" functions, but it has the additional complication of having randomness in the function evaluation. As our computational capacity and algorithmic knowledge increases, we can apply global optimization to real world problems that previously were not even considered to be framed in optimization terms. Also, as the applications in global optimization develop, they will motivate new and improved methods. Thus there is a synergy between applications and algorithmic techniques.

The ultimate goal of global optimization techniques is to develop a single method that:

- works for a large class of problems,
- finds the global optima with an absolute guarantee, and
- uses very little computation.

We are very far from achieving this goal. Typically the optimization technique is chosen to match the structure of the relevant problem. For example, a practitioner solving a linear program would select a different algorithm than if the problem was nonlinear and convex. If the problem was a linear programming problem, there are many methods that take advantage of the linear structure of the problem [113]. Similarly, a convex problem which is twice continuously differentiable would be better solved with an algorithm that uses the Hessian to take advantage of that structure [13, 103]. However, if the problem was nonlinear and multimodal with mixed continuous and discrete variables, the practitioner would have difficulty selecting the best algorithm to use, and may experiment with a few methods that would then be tailored to the problem at hand. Thus, for global optimization it is still an open research question as to the best choice of algorithm given a particular problem. This indicates the need to develop a better understanding of the algorithm behavior on various problem types.

1. Classification of Optimization Problems

Figure 1.1 illustrates a categorization of optimization problems. It is similar to the NEOS optimization tree [119] in the first division on continuous and discrete variables, but then the trees differ. The intent of Figure 1.1 is to depict a hierarchy of problems. In the categorization of the figure, the first division is on the type of domain, whether the variables are continuous or discrete (a mixed variable category is not included in the figure). Later in the theoretical analysis of pure adaptive search (Chapter 2) and hesitant adaptive search (Chapter 3), similarities are drawn between continuous problems and discrete problems. Further

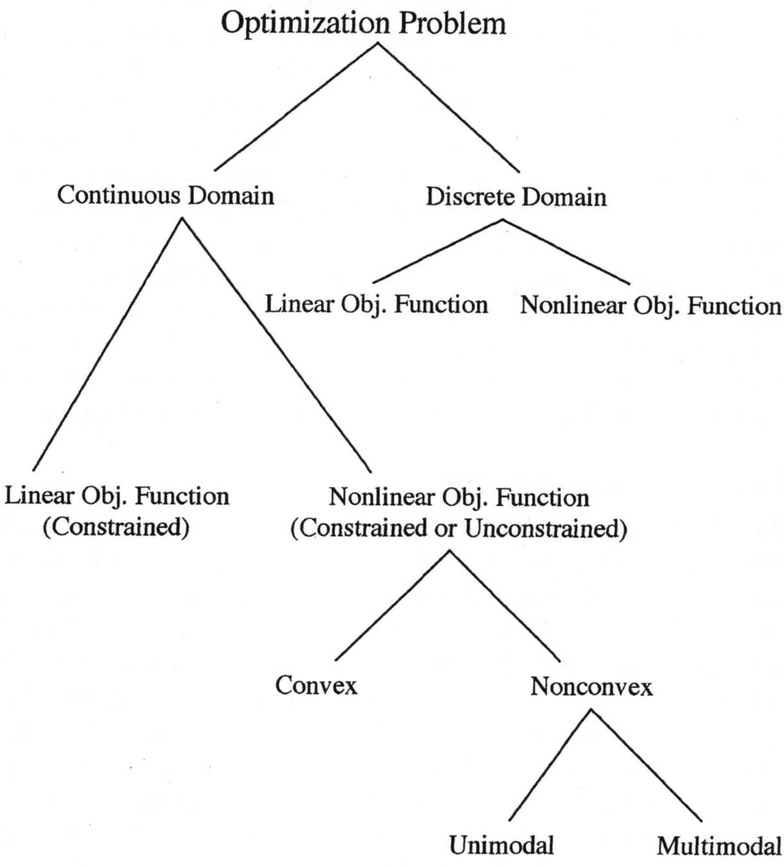

Figure 1.1. Categorization of optimization problems.

categories of the domain may include constrained (bounded) or unconstrained. When the feasible set is constrained, it could be characterized as to whether the feasible set was convex or nonconvex, and whether the equations defining the feasible set were linear or nonlinear. These characterizations are not depicted in the figure.

The second level in Figure 1.1 distinguishes between linear and nonlinear objective functions. Notice for discrete problems, the objective function distinguishes the problem as either a linear discrete or nonlinear discrete problem, whereas commonly an "integer program" presumes the objective function is linear. The characterization of nonlinear continuous problems is further broken down into convex and nonconvex, where nonconvex is typically considered *global optimization*. The nonconvex functions may be unimodal or multimodal. For example, minimizing a concave function over a set of constraints, would fall into the nonconvex multimodal category. This categorization does not rely on derivatives, and hence allows a nondifferentiable convex function (e.g., a function with breakpoints) to fall into the same category as twice continuously differentiable convex functions. On the discrete domain side, there is not an analogous concept of convexity. In fact, discrete feasible sets are never convex according to Rardin [129, page 114]. However, it may be useful to extend the definition of convexity to discrete feasible regions using an algorithmic perspective based on neighborhoods of discrete points. Discrete problems with nonlinear objective functions are considered *global optimization* problems in this book. This organization of optimization problems may suggest categories of problems that share similar characteristics and thus may guide algorithmic development.

2. Types of Algorithms

Global optimization algorithms are often classified as either deterministic or stochastic. The focus of this book is on stochastic methods that can be applied to global optimization problems with little known structure, such as "black-box" functions. There are several excellent books on global optimization, including the two volume Handbook of Global Optimization [72, 122], an overview of deterministic methods by Horst and Tuy [73], and an introduction to global optimization with stochastic methods by Törn and Žilinskas [165]. A stochastic method in this book refers to an algorithm that uses some kind of randomness (typically a pseudo-random number generator), and may be called a Monte Carlo method. Examples include pure random search, simulated annealing, and genetic algorithms.

Why stochastic search as opposed to deterministic search methods? Random search methods have been shown to have a potential to solve large problems efficiently in a way that is not possible for deterministic algorithms. Dyer and Frieze [48] showed that estimating the volume of a convex body takes an exponential number of function evaluations for *any* deterministic algorithm, but if one is willing to accept a weaker claim of being correct with an estimate that has a high probability of being correct, then a stochastic algorithm can provide such an estimate in polynomial time. Thus there is a trade-off between the amount of computation and the type of guarantee of optimality. The analyses in this book also relax the requirement of providing an absolute guarantee of the global optimum and instead are satisfied with a probabilistic estimate of the global optimum. One question is whether a stochastic algorithm can be executed in polynomial time, on the average, while it is known that a deterministic method for global optimization is NP-hard

[173]. This is at the heart of the research presented in this book, and is explored in more detail in subsequent chapters.

Another advantage to stochastic methods is that they are relatively easy to implement on complex problems. Simulated annealing, genetic algorithms, tabu search and other random search methods are being widely applied to continuous and discrete global optimization problems [129]. Because the methods typically only rely on function evaluations, rather than gradient and Hessian information, they can be coded quickly, and applied to a broad class of ill-structured problems. A disadvantage to these methods is that they are currently customized to each specific problem largely through trial and error, and there is little theory to support the quality of the solution. A common experience is that the stochastic algorithms perform well and are "robust" in the sense that they give useful information quickly for ill-structured global optimization problems.

A general theory of performance of stochastic search algorithms is presented in this book. The basic measure of performance is the number of iterations until first sampling within ϵ of the global optimum. The analysis of performance is developed by investigating a series of algorithms that are theoretical in nature - because they assume properties of the sampling distribution that may not be practically implemented. However their analysis motivates algorithms that are practical. The performance analysis also provides an understanding of how the sampling distribution of an algorithm is related to the performance measure.

3. Definitions and Assumptions

The basic global optimization problem (P), used throughout the book, is defined as,

$$(P) \qquad \min_{x \in S} f(x) \tag{1.1}$$

where x is a vector of n decision variables, S is an n-dimensional feasible region and assumed to be nonempty, and f is a real-valued function defined over S. The goal is to find a value for x contained in S that minimizes f. Let the global optimal solution to (P) be denoted by (x_*, y_*) where

$$x_* = \arg\min_{x \in S} f(x) \tag{1.2}$$

and

$$y_* = f(x_*) = \min_{x \in S} f(x). \tag{1.3}$$

It will also be convenient to define

$$y^* = \max_{x \in S} f(x).$$

In order to ensure a global optimum exists, we need to assume some regularity conditions. If (P) is a continuous problem and the feasible set S is nonempty and compact, then the Weierstrass theorem from classical analysis guarantees the existence of a global solution [103]. If (P) is a discrete problem and the feasible set S is nonempty and finite, a global solution exists. The existence of a global optimum can be guaranteed under slightly more general conditions, but these conditions are sufficient for the purposes of this book. Note that the existence of a *unique* minimum at x_* is not required. If there are multiple optimal minima, let x_* be an arbitrary fixed global minimum.

A distinction is usually made between local optima and global optima. A local optimum will be defined as a feasible point $\hat{x}$ such that sufficiently small neighborhoods surrounding $\hat{x}$ contain no points that are both feasible and improving in objective function value. For continuous domains, a small neighborhood is typically a ball of radius δ, where $\delta > 0$. For discrete domains, the set of nearest neighbors to $\hat{x}$ could be used to determine whether $\hat{x}$ is a local optimum. Notice the relationship between the concept of small neighborhood and local optimum. The definition of neighborhood, especially for discrete problems, is usually associated with an algorithm, rather than the definition of the problem, which implies that a point $\hat{x}$ might be a local optimum with respect to one algorithm and neighborhood structure, but not with respect to a different algorithm and neighborhood structure. For example, the Traveling Salesperson Problem has several possible neighborhood structures that might impose different interpretations of local optima. Thus it is important to remember that a local optimum is related to a neighborhood structure. The definition of global optima (in Equations 1.2 and 1.3) is more straightforward because it is relative to the entire feasible region, not just a local region.

The contours of the objective function of a global optimization problem in two dimensions provides a useful visualization. The level set of the function may be interpreted as the set defined by a single contour. The level set at value y, denoted

$$S(y) = \{x : x \in S \text{ and } f(x) \leq y\},$$

is the set of feasible solutions whose objective function values are y or better. The level set is defined as including points with equal objective function values, but in some algorithms we also define an improving set to only include points that are strictly improving. This distinction becomes important in the analysis of discrete problems. It is discussed in more detail when describing pure adaptive search on a finite domain, in Chapter 2. Also let $N(y)$ be the number of iterations needed to first achieve an objective function value of y or less, which corresponds to the number of iterations to first obtain a sample point in the level set $S(y)$. It is also convenient in Chapter 3 to define $M(y)$ as the number of iterations just before landing in $S(y)$, and the relationship $N(y) = 1 + M(y)$ accounts for the extra iteration to actually land in $S(y)$. The number of iterations needed to achieve an accuracy of y or less, $N(y)$, is the primary measure of performance for an algorithm in this book. It may be called the first passage time or first hitting time in the literature on stochastic processes. If y is close to the optimum, for example $y = y_* + \epsilon$ for a small positive value of ϵ, then $N(y_* + \epsilon)$ describes the number of iterations to get within ϵ of the optimum. In this book, we focus on the expected value of $N(y)$, and derive the distribution when possible.

3.1 Assumptions for Continuous Problems

Additional assumptions and restrictions on the generality of the global optimization problem (P) are made throughout this book as needed, however a brief discussion is given here.

A *continuous* global optimization problem is classified by the feasible region containing real-valued, continuous variables, $S \subset \mathbb{R}^n$. Typically, the feasible region for a continuous problem, S, is assumed to be a nonempty, compact set which is full-dimensional. The feasible region can be described by simple upper and lower bounds, $x_i^L \leq x_i \leq x_i^U$ for $i = 1, \ldots, n$ known as *box* constraints, or by more general functional constraints, $g_j(x) \leq 0$ for $j = 1, \ldots, m$. In general, S may form a nonconvex set, which can possibly be disconnected. In this case, we usually assume upper and lower bounds on the variables are known, such that S is contained in a box, or hyperrectangle.

For a continuous global optimization problem, there are several traditional ways to classify the objective function $f(x)$. The objective function $f(x)$ may be a linear function (e.g. linear programming), a nonlinear convex function, a unimodal but nonconvex function, or a multimodal function. Typically nonlinear programming assumes the objective function is nonlinear, convex and twice continuously differentiable, while global optimization often includes functions that are not differentiable everywhere, and may be discontinuous (e.g., step functions). These characteristics of the objective function may also be used to describe the constraint equations.

Another way to characterize a function of continuous variables is whether it satisfies the Lipschitz condition. A function f satisfies the *Lipschitz condition* with Lipschitz constant K, if

$$|f(x) - f(y)| \leq K\|x - y\| \tag{1.4}$$

for all x and $y \in S$, where $\|\cdot\|$ is the Euclidean norm on $\mathbb{R}^n$. A function satisfying the Lipschitz condition has a bound on the derivative, and according to Törn and Žilinskas [165, page 25], practical objective functions often have such bounds. It is less common to actually know the value of the bound. Many stochastic global optimization algorithms assume a Lipschitz constant exists, but do not use the actual value in the algorithm. This is in contrast to Lipschitz optimization which requires the value or an upper bound of the Lipschitz constant. Algorithms that require the Lipschitz constant and rely on estimates are discussed in [64]. Several analyses presented later in the book assume the objective function satisfies the Lipschitz condition.

3.2 Assumptions for Discrete Problems

A *discrete* global optimization problem is classified by the feasible region containing discrete variables. The feasible region may be described in several ways; for example S could be the set of integers between 0 and 100, which retains numerical properties, or S could be a non-ordered set such as $S = \{\text{red}, \text{yellow}, \text{blue}\}$, or the list of cities on a tour for the Traveling Salesperson Problem. Typically, the feasible region for a discrete problem, S, is assumed to be a nonempty and finite, although it is possible for S to be infinite with a bounded objective function. The concept of "dimension" is not always appropriate for a discrete global optimization problem, so it is difficult to compare a discrete domain with a continuous domain. One way to construct a comparable problem is to consider the distinct points in an n-dimensional lattice, $\{1, \ldots, k\}^n$. The number of points in the domain for the lattice is k^n, and as k gets large, the discrete domain resembles a continuous domain of dimension n. Later in this book, a lattice is used to compare a discrete problem with a continuous one.

The objective function for a discrete global optimization problem is a real-valued function defined on the points in S. The objective function $f(x)$ may have a functional form, or be evaluated by the means of a subroutine or computer code on the possible points in S. For example, $f(x)$ and S for a continuous global optimization problem can be turned into a discrete problem by simply adding the constraint that x be integer valued. The objective function for a discrete problem with integer variables may be considered linear or nonlinear with respect to the variables when they are relaxed to be continuous variables. This is interesting when algorithms (such as simulated annealing) define neighborhoods that may not be a natural neighborhood on the relaxed problem. Analogous definitions of linearity, convexity, Lipschitz condition, and other concepts must be extended to discrete problems to better generalize methods that are appropriate for both continuous and discrete domains.

3.3 Mixed Continuous-discrete Problems

A *mixed continuous-discrete* global optimization problem is classified by the feasible region containing both continuous and discrete variables. In engineering applications, it is often convenient to define a global optimization problem for continuous variables, and then explore the effects of limiting a subset of the variables to discrete values. In Chapter 7, a 10-bar truss problem is described, where the decision variables are the diameters of the bars. While the diameter of a bar is mathematically a continuous variable, in reality, bars are only manufactured and readily available with a discrete set of standard diameters. Thus a continuous problem becomes a discrete one when taking practical considerations into account.

Another example, described in more detail in Chapter 7, is a composite structure, where the number of plies is an integer variable while the height and width of stiffeners are continuous variables. Additional variables in a composite structure are the fiber angles of the plies. The fiber angle of a ply is similar to the diameter of a truss member in that mathematically the angle could take on any value between ± 90 degrees, however manufacturing technological constraints restrict the angle to a discrete set of values, e.g., 0, ± 45, or ± 90 degrees. In order to solve these types of real-world problems, we need robust algorithms that can be applied to global optimization problems with a mixture of continuous and discrete variables.

4. Overview of Random Search Methods

One motivation for random search methods in global optimization is the potential to obtain approximate solutions quickly and easily. In addition, theoretical analysis of random search methods indicates that performance may be very good, possibly polynomial in dimension, despite the fact that global optimization problems are NP-hard for deterministic methods. A brief overview of random search methods is presented in this section to provide some context for stochastic methods for global optimization. Underlying all of these methods is a probabilistic approach to sampling the feasible region. Whether the algorithm is simulated annealing, a genetic algorithm or multistart, it has some method of generating new candidate points, which can be called its sampling distribution. The sampling distribution employed by the algorithm is used in the subsequent analyses to characterize the performance of the method.

We start with a brief description of exhaustive search, including grid search and pure random search. Then we present a framework for sequential random search and a brief discussion of simulated annealing. This is followed by a framework for two-phase methods, including multi-start, clustering, single linkage and multi-level single linkage algorithms. Finally a brief overview of population-based algorithms is discussed, including a framework for genetic algorithms and similar methods. The common theme carried throughout the book is studying the probability distribution of the points generated by an algorithm, and the impact the sampling distribution has on the complexity and behavior of the algorithm on classes of global optimization problems.

4.1 Enumeration or Exhaustive Search

When confronted with a global optimization problem, the basic and perhaps most natural approach is to simply evaluate all points in the domain. If the domain S is finite and relatively small, an exhaustive search is a reasonable approach. As Rardin notes [129, page 627], if the domain has only a few discrete decision variables, the most effective optimization method is often the most direct: enumeration of all the possibilities. However, in most discrete global optimization, it is not practical to perform complete enumeration. Combinatorial optimization includes the Traveling Salesperson Problem, which for an N-city tour, has $(N-1)!$ points in the domain. For $N = 10,000$, this is over 10^{20} possible tours, and at one CPU second per function evaluation, brute force enumeration would exceed 10^{12} years.

Grid Search. When the global optimization problem involves continuous variables, there are an infinite number of points in the domain, and complete enumeration is impossible. A common approach is to perform a *grid search,* essentially discretizing the domain. A grid search creates an equally spaced grid of points over the feasible region, and evaluates the objective function at each point. If the objective function satisfies the Lipschitz condition with constant K and the domain is an n-dimensional hyperrectangle of maximum length D on each side, then the grid spacing can be determined (see Dixon and Szegö [44, 45]) to achieve a desired accuracy, of having the estimate within ϵ of the global optimum. If the spacing between grid points is ϵ/K on each coordinate, then there are approximately $(KD/\epsilon)^n$ grid points, and the possible error between adjacent points is bounded by ϵ. Hence, the number of evaluations needed to obtain an accuracy of ϵ is proportional to $(KD/\epsilon)^n$. Thus, the number of function evaluations to achieve an accuracy of ϵ is exponential in the dimension n.

Pure Random Search. A stochastic version of grid search is pure random search. *Pure random search,* discussed in Chapter 2, was first defined by Brooks [26], discussed by Anderssen [6], and later named in the classic volumes by Dixon and Szegö [44, 45]. Pure random search samples repeatedly from the feasible region S, typically according to a uniform sampling distribution. Although the points of pure random search are not evenly spaced, as in grid search, they are uniformly scattered over the feasible region. If the objective function has some regularity that coincides with the regularly spaced grid points (e.g., a sinusoidal function), then the probabilistic nature of pure random search provides an advantage. While pure random search sacrifices the guarantee of determining the optimal solution within ϵ, it can be shown that pure random search converges to the global optimum with probability one. A similarity between grid search and pure random search is that the expected number of function evaluations for pure random search to get within ϵ is also exponential in the dimension n (see Chapter 2).

Other Covering Methods. Other sampling methods that eventually cover the feasible set may also be used in a type of exhaustive search. Törn and Žilinskas discuss other covering methods [165, Chapter 2] which provide a means to sample points throughout the feasible region in a thorough manner. They include quasi-random sequences, such as introduced by Halton (see [165, page 33], or [63]) as providing a uniform covering. Unfortunately, the rate of convergence for a covering method is in general slow, as is demonstrated by the exponential complexity of both grid search and pure random search. To improve performance, some means to focus the search on promising regions is needed.

4.2 Sequential Random Search

Stochastic algorithms tend to be categorized into two classes, sequential algorithms, and two-phase methods which include multistart and clustering methods. Section 4.2 provides an overview of sequential algorithms, including simulated annealing, and Section 4.3 provides a brief description of two-phase methods. It should be noted that this categorization is blurred as new algorithms are being developed. For example, it is not clear which categorization includes genetic algorithms; here they are summarized in Section 4.4.

Sequential random search, including simulated annealing, has been applied to many "black box" global optimization problems. Sequential random search procedures can be characterized by producing a sequence of random points $\{X_k\}$ on iteration $k, k = 0, 1, \ldots$ which may depend on the previous point or several of the previous points. We provide a framework for sequential random search.

Sequential Random Search

Step 0. Initialize algorithm parameters and initial point $X_0 \in S$ and set iteration index $k = 0$.

Step 1. Generate a candidate point $V_{k+1} \in S$ according to a specific *generator*.

Step 2. Update the current point X_{k+1} based on the candidate point and previous points.

Step 3. If a stopping criterion is met, stop. Otherwise update algorithm parameters, increment k and return to Step 1.

Sequential random search depends on two basic iterations, the generator in Step 1 that produces candidate points, and the update procedure in Step 2 that may accept a candidate point to be the next point in the sequence. The sequence of points X_k provides a search path through the feasible set. If the objective function is consistently improving, $f(X_0) > f(X_1) > \cdots$ then the algorithm has an improving search. If, as in simulated annealing, non-improving points are occasionally accepted in the update procedure, the search path is not consistently improving. In all practical implementations that this author has seen, a part of Step 3 includes recording the best point found so far. In this way, an algorithm may update X_{k+1} with a non-improving point without risk of losing the value of the incumbent solution. This has implications when discussing convergence of an algorithm, and distinguishing whether the algorithm converges to the global optimum with respect to the current point, or the algorithm converges to the global optimum with respect to the best point sampled thus far.

Simulated Annealing. Simulated annealing may be viewed as a type of sequential search algorithm. Simulated annealing was motivated by the physical annealing process when slowly cooling metals, and introduced by Metropolis, et al. [110], and later by Kirkpatrick in 1983 [89]. Simulated annealing has most often been applied to combinatorial problems (see Aarts and Korst [1]), such as the Traveling Salesperson

Problem [89] and various scheduling problems [126], but has also been applied to continuous or mixed domain problems [18, 36, 140, 141, 168]. While simulated annealing had a burst of popularity in the 80's and 90's, other sequential random search algorithms predate it. Several of these sequential random search methods were reported as experiencing computation that is linear in dimension, and are discussed later in this section.

The generator, or method of generating candidate points as in Step 1, is usually specific to each algorithm and tailored for the problem at hand. The generator usually makes a move in the vicinity of the current point X_k on iteration k. In analyses, the generator is often viewed as a Markov chain characterized by the one-step transition probability. In continuous domains, the generator often involves making a step in a specified direction. Then the transition probability would correspond to the probability of selecting a particular direction and step. In discrete domains, a similar step generator may be used, however it is more common for discrete generators to be described as selecting a candidate point at random from a neighborhood, where the definition of the neighborhood is specific to the problem. A discrete neighborhood may be the set of nearest neighbors, that is the set of points that can be sampled with one transition. For the Traveling Salesperson Problem, several generators have been proposed, including a one-city swap, or a k-city swap where k links are deleted and replaced in a different way that maintains feasibility [46]. A generator based on Hit-and-Run is discussed in Chapter 6 that does not have to be tailored for specific problems, but can be defined for a general class of global optimization problems.

A typical feature of simulated annealing distinguishing it from other sequential random search algorithms is that, besides accepting points that have improvements in objective function value, it also has a probability of accepting non-improving points. The update procedure in Step 2 of sequential random search for simulated annealing is

$$X_{k+1} = \begin{cases} V_{k+1} & \text{with acceptance probability } P_{T_k}(X_k, V_{k+1}) \\ X_k & \text{otherwise} \end{cases}$$

for any iteration k. The parameter T_k is referred to as the temperature, and it is initialized at a large value. The temperature controls the probability of accepting a non-improving point; when the temperature is high there is a large probability of accepting a non-improving point, and as the temperature decreases to zero the probability of accepting a non-improving point also decreases to zero.

The acceptance probability P_{T_k} of accepting a candidate point V_{k+1}, given the current iteration point X_k and the current temperature T_k, is typically given by,

$$P_{T_k}(X_k, V_{k+1}) = \begin{cases} 1 & \text{if } \textit{improving}, \\ & \text{i.e., } f(V_{k+1}) < f(X_k) \\ e^{\left[\frac{f(X_k)-f(V_{k+1})}{T_k}\right]} & \text{otherwise}, \\ & \text{i.e., } f(V_{k+1}) \geq f(X_k) \end{cases} \tag{1.5}$$

The acceptance criterion based on P_{T_k} is also known as the Metropolis criterion [110]. The Metropolis criterion was introduced to model the cooling of metals, and the difference in objective function values $f(X_k) - f(V_{k+1})$ was referred to as the energy difference, and the temperature T_k involved the temperature of the heat bath as well as a physical constant known as the Boltzmann constant. The temperature is gradually reduced, and if the lowering of the temperature is sufficiently slow, the metal can reach thermal equilibrium at each temperature. The sequence of points generated while holding temperature constant T converges to the Boltzmann distribution, which characterizes the thermal equilibrium (page 14, [1]). The Boltzmann distribution plays an important role in characterizing the underlying probability of the points sampled and accepted. This is discussed further in Chapter 4.

The rate at which temperature is gradually reduced is critical to the annealing process, and if lowered too quickly results in an inferior metal. The optimization analogy is that if the temperature is lowered too quickly, the algorithm gets trapped at a local optimum. The control mechanism to reducing the temperature is called the cooling schedule. Let $\tau(T_k)$ be a function that gradually reduces the temperature T_k on iteration k. A simple geometric cooling schedule is $\tau(T_k) = 0.9T_k$. Initially when the temperature T_k is large, many non-improving points will be accepted, but as T_k approaches zero, mostly improving points will be accepted. This feature allows the algorithm to escape from local optima when T_k is large, but allows convergence to, hopefully the global optimum, as T_k becomes small. A cooling schedule often attempts to allow the optimization process to reach thermal equilibrium (e.g., the Boltzmann distribution) by lowering the temperature after a minimum number of iterations N_{T_k} at each temperature step. Often the cooling schedule is fine-tuned for a particular problem, but recent results derive an analytically motivated cooling schedule [86, 140]. Some of these results are summarized in Chapter 4.

We now summarize the simulated annealing algorithm.

Simulated Annealing

Step 0. Initialize algorithm parameters, including temperature T_0 and initial point $X_0 \in S$ and set iteration index $k = 0$.

Step 1. Generate a candidate point $V_{k+1} \in S$ according to a specific *generator*.

Step 2. Update the current point X_{k+1} using

$$X_{k+1} = \begin{cases} V_{k+1} & \text{with probability } P_{T_k}(X_k, V_{k+1}) \\ X_k & \text{otherwise} \end{cases}$$

where $P_{T_k}(X_k, V_{k+1})$ is given in Equation 1.5. Update algorithm parameters, including $T_{k+1} = \tau(T_k)$.

Step 3. If a stopping criterion is met, stop. Otherwise increment k and return to Step 1.

The convergence properties of simulated annealing have been analyzed in [1] for combinatorial optimization problems. The algorithm is analyzed using Markov chain theory, where iterations correspond to transitions and the state space is the finite set of outcomes. It is shown that simulated annealing asymptotically converges to the set of global optimal solutions with probability one. This is done by proving that the Markov chain describing the algorithm converges to a stationary distribution. The number of transitions required to approximate the stationary distribution depends on the second largest eigenvalue of the transition matrix. This can be used to show that the stationary distribution is approximated arbitrarily closely, only if the number of transitions is at least quadratic in the size of the solution space. For instance if the solution space S is an n dimensional binary lattice, then there are 2^n possible solutions in the solution space where n denotes the dimension of the problem. Therefore it will take at least $(2^n)^2$ transitions to verify the global optimal solution. This means that approximating the stationary distribution arbitrarily closely results in an exponential time execution of the simulated annealing algorithm. Notice that this involves verifying the global optimal solution. The authors do not give the expected number of iterations to find the global optimal solution for the first time. Other analyses by Locatelli [99, 100] and Trouvé [167] provide conditions for which simulated annealing converges in probability to the global optimum. Romeijn, et al. [141] provide convergence results for simulated annealing with both continuous and discrete variables. This is discussed in more detail in Chapter 4.

Step Size Algorithms. A common method to generate a candidate point on problems with continuous variables is take a step size in a vector direction, called a direction-step paradigm in [129]. In continuous problems, the direction of movement may be based on gradient information, although not necessarily. Sequential random search, in Step 1, typically generates a candidate point by taking a step of length S_k in a specified direction D_k:

$$X_{k+1} = X_k + S_k D_k$$

on an iteration k. In a gradient search type algorithm, the direction is based on local information by evaluating the gradient at the current point, and the step length may be the result of a line search. Quasi-Newton methods take advantage of an approximation of the Hessian to provide a search direction. As an alternative, a direction is often generated according to a distribution (often uniform on a hypersphere), and the step length may also be randomly generated.

A collection of sequential step size algorithms, including those of Rastrigin, et al. [114, 131, 132], Steiglitz, et al. [97, 152], Schrack, et al. [150, 151], and Solis and Wets [159] fit into this category of sequential random search, and all obtain a direction vector by sampling from a uniform distribution on a unit hypersphere. The method of choosing the is specific to each algorithm, such as shrinking or expanding the step length based on previously chosen points. Several of these sequential random search methods have been reported as experiencing computation that is linear in dimension.

After a candidate point is generated, Step 2 of sequential random search specifies a procedure to update the current point. Algorithms that are strictly improving have a simple procedure, update the current point only if the candidate point is improving,

$$X_{k+1} = \begin{cases} V_{k+1} & \text{if } f(V_{k+1}) < f(X_k) \\ X_k & \text{otherwise.} \end{cases}$$

This type of improving algorithm may get trapped in a local optimum if the generator does not sample over the entire domain. If the neighborhood, or procedure for generating candidate points is too restricted, it is difficult to find the global optimum. One remedy is to sample a large neighborhood, possibly draw from the entire feasible set, and another remedy is to accept non-improving points (as in simulated annealing, Section 4.2). The algorithms based on Hit-and-Run sampling methods presented in Chapters 2, 3 and 6 use a global reaching search strategy so there is a positive probability of sampling anywhere in the entire feasible region. Recent research in Very Large Scale Neighborhood Search is pursuing the benefit of enlarging the neighborhood search. The computational tradeoffs between sampling over a very large neighborhood and possibly the entire feasible region, versus sampling over a restricted neighborhood but accepting non-improving points is explored in subsequent chapters.

Convergence. A convergence proof for sequential random search algorithms was provided by Solis and Wets [159], where convergence means that, with probability 1, the sequence $f(X_k)$ converges to the infimum of f on S as the iteration counter k tends towards infinity. The convergence theorem for global search [159, page 20] makes two assumptions. The first assumption is roughly that the update procedure (Step 2 in the sequential random search algorithm) chooses the best of the points found thus far, and the second, more restrictive assumption, is that given any subset A of S with positive "volume," the probability of repeatedly missing the set A when generating the random samples (in Step 1) must be zero. It basically says that, as long as the method of generating a subsequent point does not consistently ignore any region, then the algorithm will converge with probability one. This is particularly relevant to the step-size algorithms of Rastrigin, Schumer and Steiglitz, and others [97, 131, 132, 114, 152, 150, 151].

A second convergence proof due to Bélisle [14] says that, even if the generator of an algorithm cannot reach any point in the domain in one iteration, if there is a means such as an acceptance probability to allow the algorithm to reach any point in a finite number of iterations, then the algorithm still converges with probability one to the global optimum.

Algorithms for global optimization must either sample the entire set or use the structure of the problem to guarantee convergence. Stephens and Baritompa [161] formalize this by proving that convergence requires global information. Examples of global information that may improve performance of an algorithm include the Lipschitz constant, bounds on derivatives, bounds on the function as in interval methods, information on the level sets, number of local optima, functional form, and the global optimum itself. Stephens and Baritompa show that deterministic algorithms must sample a dense set to find the global optimum value, and proved analogous results for stochastic algorithms. They show [161, Theorem 3.2] that for any deterministic sequential sampling algorithm on a sufficiently rich class of functions F, there exists a function in F for which the algorithm fails to detect the global optimum. The analogous stochastic result [161, Theorem 3.4] is that, for any stochastic sequential sampling algorithm and any $\epsilon > 0$, there exists a function in F such that the probability that the algorithm detects the global optimum is less than ϵ. While these results show that attempts to find the global optima on *all* functions is doomed to failure, Stephens and Baritompa conclude that using "global optimization heuristics is often far more practical than running general algorithms until the mathematically proven stopping criteria are satisfied" [161, page 587]. They point to the need to quantify the 'niceness' of realistic functions so that practical problems and algorithms may be combined with mathematical confidence in the results. By relaxing the criteria of guaranteeing a global solution to a weaker claim of probabilistically detecting the global optimum, the analyses in the book hope to be useful in bridging the gap between mathematical confidence and practicality.

Why is it that sometimes random search algorithms appear to find a global optimum quickly, when other times they appear to get trapped at a local optimum? The rate of convergence is one way to characterize performance. With random search algorithms, the measure of performance should include both the speed of the algorithm as well as the accuracy of the final solution. This is explored in subsequent chapters.

There is experimental evidence in the literature that suggests sequential random search algorithms are efficient for large dimensional quadratic programs. Schumer and Steiglitz [152] provide experimental evidence that the number of function evaluations increases linearly with dimension for their adaptive step size algorithm on the following three test functions: $\sum_{i=1}^n x_i^2$, $\sum_{i=1}^n x_i^4$, and $\sum_{i=1}^n a_i x_i^2$. They also prove that the average number of function evaluations for an optimum relative step size random search restricted to an unconstrained quadratic objective function is asymptotically linear in dimension. Schrack and Borowski [150] report experimental results on a quadratic test function, $\sum_{i=1}^n x_i^2$, that doubling the dimension doubles the number of function evaluations required for their random search algorithm. Solis and Wets [159] experimentally verified a linear correlation between the number of function evaluations and dimension for their own variation of the step size algorithm on a quadratic test function. They provided a justification of this linearity condition based on the tendency of these algorithms to maintain a constant probability of successful improvement.

There is also theoretical justification that sequential random search algorithms are efficient for a larger class of global optimization problems. In Zabinsky and Smith [188], complexity is measured as the expected number of iterations needed to get arbitrarily close to the solution with a specified degree of certainty, and this measure is used throughout the book. An analysis of a random search procedure called pure adaptive search [125, 188] proves that it is theoretically possible for a sequential random search procedure to achieve linear complexity (in improving iterates) for global optimization problems satisfying the Lipschitz condition. If sequential random search algorithms behave similarly to pure adaptive search, the analysis would explain why they appear efficient. The linearity performance of pure adaptive search is discussed in detail in Chapter 2.

4.3 Two-Phase Methods

Many global optimization algorithms may be thought of as having two phases, a global phase when sampling occurs over the entire feasible region, and a local phase when sampling occurs over a restricted, or focused, portion of the feasible region. A common example of combining a global phase with a local phase is a multistart method. Multistart samples starting points (often uniformly) from the entire set during its global phase, and uses them to initiate deterministic gradient search type algorithms as a part of its local phase. It is also useful to consider other algorithms as a part of a global and local phase when considering the performance or behavior of the algorithms. For instance, a multistart scheme may be used where simulated annealing can be viewed as the local phase. The global phase can be viewed as an exploratory phase aimed at exploring the entire feasible region, while the local phase can be viewed as an exploitation phase aimed at exploiting the location and/or local information (e.g. gradient) to improve on the objective function.

Schoen [149] provides a general scheme for a two-phase method, and Wood, et al. [178] also states a generic stochastic optimization algorithm. A generic two-phase algorithm is stated as follows.

Basic Two-Phase Stochastic Global Optimization Algorithm

Step 0. Initialize algorithm parameters and set iteration index $k = 0$.

Step 1. In the global phase, generate $X_k \in S$ according to a sampling distribution over S.

Step 2. In the local phase, generate a candidate point $V_k \in S$ according to a local sampling distribution, or local descent algorithm, and update information on the best solution found so far as well as other parameters of interest.

Step 3. If a stopping criterion is met, stop. Otherwise update algorithm parameters, increment k and return to Step 1.

The multistart algorithm fits into this framework, where the global phase generates a point using a sampling distribution, and the local phase performs a type of local search. Typically the local phase in multistart is a deterministic gradient search type algorithm, although experiments using simulated annealing and other sequential random search methods within multistart have promising results [42]. A multistart algorithm may repeatedly find a local optimum from many starting points, and clustering methods have attempted to modify the multistart idea by reducing the number of times a local search is initiated. Clusters are formed (often grown around a seed point) to predict whether a starting point is likely to lead to a local optimum already discovered, and local searches are only initiated at promising candidate points (see [165, 149]). A related idea has led to linkage methods, which "link" points in the sample and essentially view clusters as trees, instead of spherical or ellipsoidal clusters. The most well known linkage method is Multi Level Single Linkage [137, 138] and several variants are summarized in [149].

4.4 Genetic Algorithms

The division between the global phase and the local phase blurs when combining algorithms. For instance, when a simulated annealing is coupled with multistart the local phase has a global aspect to it. Population-based algorithms, including genetic algorithms and evolutionary programming [91], maintain a set of current points called the population. The current population is used to generate candidate points for a new population, typically by combining pairs of points with specific rules of crossover, mutation and reproduction. The current and new points are then evaluated and compared to update the population. The motivation behind population-based algorithms is to parallel the process of biological evolution.

Several details must be specified to completely define the algorithm. First, the population size must be chosen. The *elitist strategy* subdivides the population into three categories and maintains p_e elite (best) solutions, p_i immigrant solutions (added to maintain diversity), and p_c crossover solutions. So the total population size is $p = p_e + p_i + p_c$ (described in [129]). Computational experience indicates that the population size impacts performance, if it is too small, the algorithm has difficulty finding the global optimum, and if it is too large, then it is inefficient and essentially pure random search. The right population size for a particular problem is usually found through experimentation.

Given a current population of p points, the next step is to generate candidate points to use in a new population. The primary mechanisms are crossover, mutation, and reproduction. Crossover implies that two "parent" solutions in a population are used to generate two "children" solutions by breaking both parent solutions and reassembling the first part of one with the second part of the other and visa versa. This is usually done with binary variables, but variations on continuous variables have also been demonstrated. It appears that direct crossover on continuous variables may be more efficient than encoding a real-valued number into binary and then performing crossover. Mutation is used to randomly alter a single solution. Reproduction is used to duplicate promising solutions which modifies the composition of the population.

The final step is to merge the current population with the new candidate points. A naive approach would be to simply rank order the population on objective function values and select the best, however experience has shown that some diversity should be maintained, to prevent premature convergence to a local optimum. Various fitness measures have been proposed, and different strategies to merge the current and new populations by selecting "survivors" and "immigrants" to maintain the population size. As with simulated annealing, the best solution is always recorded, but unlike simulated annealing, a population of alternative solutions is provided. Of course, practitioners could keep track of alternative solutions in a simulated annealing algorithm too, and keep a list of the p best solutions if that is of interest.

In keeping with the theme of the book, the method of generating candidate points may be viewed as a probability distribution. Researchers have used Markov chain analysis [40] to analyze the behavior, such as the expected waiting time, of genetic algorithms [92]. This is similar to analyses presented in subsequent chapters.

4.5 Other Stochastic Methods

There are many other stochastic methods that have been proposed for global optimization with discrete and/or continuous variables that will not be mentioned in this book. Some interesting methods include, tabu search [55], nested partitioning method [156], Lipschitz optimization [127], controlled random search [4], localization search [11, 177], and uniform covering using raspberries [69]. There are even more deterministic methods, and hybrid algorithms combining deterministic and stochastic methods (e.g. [93]). This book is not able to review them all, but instead strives to present a uniform way of viewing stochastic algorithms.

5. Overview of this Book

In Chapter 2, we present a theoretical development of the performance of two stochastic methods that can be viewed as extremes. These two random search algorithms, pure random search (PRS) and pure adaptive search (PAS), are not intended to be practical algorithms, but rather are used to describe the performance of random search algorithms based on an underlying sampling distribution. It is shown that, under certain conditions, the expected number of iterations of pure adaptive search is *linear* in dimension, while pure random search is exponential in dimension.

Chapters 3, 4 and 5 discuss relaxations of PAS: hesitant adaptive search, annealing adaptive search, and backtracking adaptive search. Hesitant adaptive search is a generalization of PAS that includes an expression of the extra computation associated with not being able to always generate improving points. Adaptive search is a relaxation of PAS by changing the sampling distribution in a way that maintains the linearity result for improving points, and is discussed in Chapter 4. Backtracking adaptive search extends the analysis to include a sequence of points that include non-improving points, backtracking in terms of the objective function. This concludes the analyses of theoretical algorithms in this book.

Chapter 6 presents an initial attempt to realize PAS by generating points that are approximately uniform. The Improving Hit-and-Run algorithm is defined in Section 2, and is shown to have a polynomial computation on average for the class of positive definite quadratic programs. Section 3 presents the Hide-and-Seek algorithm, which is motivated by Adaptive Search in the same manner that Improving Hit-and-Run was motivated by Pure Adaptive Search. Both algorithms are based on the Hit-and-Run generator, and variations are summarized later in the chapter, as well as initial results on a discrete form of Hit-and-Run. Finally, several engineering design applications, particularly the optimal design of composite structures, are discussed in Chapter 7.

6. Summary

Engineers using gradient-based local optimization methods have commented that their problems tend to have many local optima [62, 185]. The most common global optimization method found in practice is multistart, a combination of random starting points and local searches [62]. However newer methods are quite powerful, and with the advances in computer technology, are becoming practical. Simulated annealing and genetic algorithms are useful for new ill-structured applications. The primary structure these random search methods have in common is a probabilistic way of sampling the next point. Characterizing performance based on the sampling distribution is a step towards better understanding the algorithms. This can assist practitioners in selecting an appropriate algorithm for their problem, as well as developing new applications and new algorithms. Hopefully, this book and others like it can help bridge the gap between the advances in global optimization and the practical needs of the engineers.

Chapter 2

PURE RANDOM SEARCH AND PURE ADAPTIVE SEARCH

A theoretical analysis of the performance of two stochastic methods is presented, pure random search (PRS) and pure adaptive search (PAS). These two random search algorithms can be viewed as extremes and are not intended to be practical algorithms, but rather are used to provide insight and bracket the performance of stochastic algorithms. Pure random search samples points from the domain independently, and the objective function has no impact on the technique of generating the next sample point. In contrast, pure adaptive search samples the next point from the subset of the domain with strictly superior objective function values. Pure random search is extreme in the sense that the iterates are completely independent and never use previous information to affect the search strategy, while pure adaptive search is extreme in an opposite sense because the iterates completely depend on each other, and the search strategy forces improvement by definition. These two extreme stochastic methods are analyzed and it is shown that, under certain conditions, the expected number of iterations of pure adaptive search is *linear* in dimension, while pure random search is exponential in dimension. The linearity result for pure adaptive search implies that adapting the search to sample improving points is very powerful.

1. Pure Random Search (PRS)

Pure random search, first defined by Brooks [26], discussed by Anderssen [6] and later named in the classic volumes by Dixon and Szegö [44, 45] is, in some sense, the simplest stochastic method. As in Chapter 1, consider the global optimization problem (P) $\min_{x \in S} f(x)$. Pure random search generates a sequence of independent, identically distributed points in the feasible region S. Usually the points are generated according to a uniform distribution, however any probability distribution could be used. When a stopping criterion is met, the best point of the sequence generated thus far is used as an approximation to the optimal solution.

Pure random search is stated more formally below.

Pure Random Search (PRS)

Step 0. Initialize $X_0 \in S$ according to probability measure δ on S. Set iteration index $k = 0$. Set $Y_{\text{Best}} = Y_0 = f(X_0)$.

Step 1. Generate X_{k+1} according to probability measure δ on S. Set $Y_{k+1} = f(X_{k+1})$. Update the best point so far, $Y_{Best} = \min\{Y_{Best}, Y_{k+1}\}$.

Step 2. If a stopping criterion is met, stop. Otherwise increment k and return to Step 1.

To characterize the performance of PRS, consider the points $\{X_k, k = 0, 1, \ldots\}$ generated by PRS on the global optimization problem (P) as a stochastic process. These points X_k generated by PRS are independent and identically distributed over S for $k = 0, 1, \ldots$, with objective function values satisfying $Y_k = f(X_k)$. Also consider the sequence of objective function values $\{Y_k, k = 0, 1, \ldots\}$ as a stochastic process.

The underlying generating distribution for PRS may be any well-defined probability measure δ, on the feasible region S, with the restriction that any neighborhood containing the global optimum has positive measure. It is often easier to visualize the method when a common probability distribution, such as a uniform distribution, is used. Let ν represent Lebesgue measure. When pure random search uses a uniform distribution, $\delta = \nu$. The discussion of hesitant adaptive search in Chapter 3 presents a general treatment using an underlying sampling distribution δ.

To better understand pure random search, consider an example where there are ten discrete points in the domain $S = \{1, 2, \ldots, 10\}$, and $f(x)$ is given in Table 2.1. In this example, the global minimum is at $x_* = 3$ with $f_* = 1$.

The probability distribution of the objective function values, imposed by the generating distribution, is critical to the performance of the algorithm. Let $p(y)$ be the probability that a point in the domain generated by PRS falls within the level set $S(y)$, where

$$S(y) = \{x : x \in S \text{ and } f(x) \leq y\} \tag{2.1}$$

x	$f(x)$
1	3
2	4
3	1
4	5
5	6
6	2
7	3
8	5
9	6
10	3

Table 2.1. Discrete example problem with ten points in the domain, with $x_* = 3$ and $y_* = f(3) = 1$.

for $k = 0, 1, \ldots$ and $y_* \leq y \leq y^*$. Then the probability of sampling a point in the level set is the probability a point has an objective function value less than or equal to y, that is,

$$\begin{aligned} p(y) &= P(X_k \in S(y)) \\ &= \delta(S(y)) \\ &= P(f(X_k) \leq y) \\ &= P(Y_k \leq y) \end{aligned}$$

for $k = 0, 1, \ldots$ and $y_* \leq y \leq y^*$. Notice that $p(y)$ is the same for all iterations k since PRS generates identically distributed points. When the uniform distribution is employed in PRS, we have

$$p(y) = \nu(S(y))/\nu(S)$$

where $\nu(\cdot)$ denotes Lebesgue measure.

Considering the example in Table 2.1 with uniform sampling, $p(3) = 5/10$ because there are five points in the domain with objective function values of 3 or less. Similarly, the probability that the global minimum is sampled on any particular sample is $p(y_*) = 1/10$. A possible realization of PRS corresponding to this example is illustrated in Figure 2.1. The initial point sampled has an objective function value of 6, and each iteration is independently generated. The example illustrates that the objective function values are not always decreasing.

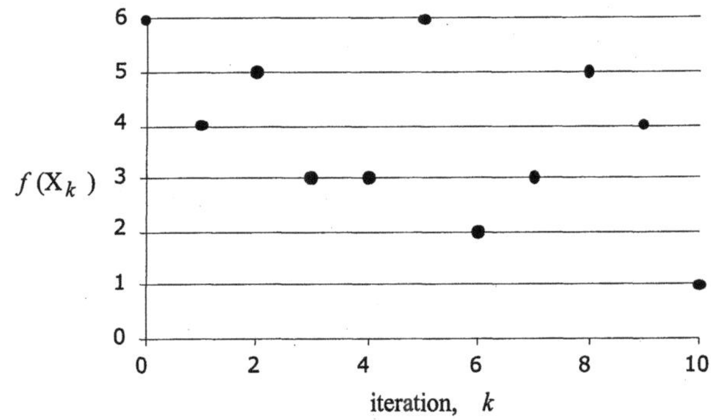

Figure 2.1. Illustration of pure random search.

In order to describe the performance of PRS, we use the number of iterations until the global minimum is first sampled as a measure of computational complexity. This measure does not include the effort reflected by a stopping criterion. It also does not include any computation needed to confirm the solution, which is a significant distinction from the definition of computational complexity used in the NP-completeness literature [173]. For PRS, an iteration performs exactly one function evaluation, so this captures the majority of the computational effort. The number of iterations until first sampling the global minimum is used throughout as an overall measure of computational complexity. It provides a general effectiveness measure, and is used to compare and contrast specific random search algorithms.

Let $N(y_*)$ be the number of iterations to first achieve a point within the level set $S(y)$, i.e.,

$$\begin{aligned} N(y) &= \min\{k : X_k \in S(y)\} \\ &= \min\{k : Y_k \leq y\} \end{aligned}$$

for $y_* \leq y \leq y^*$. Then $N(y_* + \epsilon)$ for $\epsilon > 0$ is the the number of iterations until sampling within ϵ of the global minimum y_*. This measure of complexity $N(y)$ is governed by the underlying sampling distribution. We can characterize the expected value and variance of $N(y)$ for PRS by $p(y)$. To determine the mean, consider the number of iterations until first landing in $S(y)$ as a geometric distribution [43], where the probability of a "success" is $p(y)$, the probability of sampling a point in $S(y)$. Then, for a finite discrete domain, the expected number of iterations until a point is first sampled with an objective function value of y or less is

$$E[N(y)] = 1/p(y). \tag{2.2}$$

This relationship supports the natural intuition that as it gets harder to sample the global minimum (i.e., $p(y_*)$ gets small), the expected number of iterations gets larger (inversely proportional). For the example in Table 2.1 where $p(y_*) = 0.1$, the expected number of iterations to first sample the global optimum $E[N(y_*)]$ is ten.

It is also possible to provide the variance of the number of iterations a point in $S(y)$ is first sampled in terms of $p(y)$, as

$$\mathrm{Var}(N(y)) = \frac{1 - p(y)}{p(y)^2}.$$

This supports numerical observations that pure random search experiences large variation, where the standard deviation is roughly inversely proportional to $p(y)$. As the probability of sampling the global minimum decreases, the variance of $N(y_*)$ increases.

For a continuous domain problem, the probability of sampling any distinct point is zero, so instead of being interested in $N(y_*)$, we consider the number of iterations to sample within ϵ of the global minimum, $N(y_* + \epsilon)$, where $\epsilon > 0$. Again, the number of iterations until the first sample point falls within ϵ of y_* can be thought of as a geometric distribution, and

$$E[N(y_* + \epsilon)] = 1/p(y_* + \epsilon).$$

One way to gain insight into the performance of PRS is to consider a global optimization problem where the domain is an n-dimensional ball of radius 1, and the area within ϵ of the global optimum is a ball of radius ϵ, for $0 < \epsilon \leq 1$. Using a uniform sampling distribution on this problem, $p(y_* + \epsilon) = \epsilon^n$ for $0 < \epsilon \leq 1$ and the expected number of iterations until a sample point falls within the ϵ-ball is $(1/\epsilon)^n$, an exponential function of dimension. This example also shows the effect on computation as a function of ϵ. To expand this example, suppose the objective function satisfies the Lipschitz condition (Equation 1.4) with a Lipschitz constant of K. If a sample point falls within an n-dimensional ball of radius ϵ/K about y_*, where $0 < \epsilon \leq K$, then the objective function value is within ϵ of the global minimum. The probability of sampling within this ball of radius ϵ/K, using a uniform distribution, is

$$p(y_* + \epsilon) = (\epsilon/K)^n$$

which yields

$$E[N(y_* + \epsilon)] = (K/\epsilon)^n.$$

For this situation, the expression for the expected number of iterations indicates the impact of the Lipschitz constant, as well as showing an exponential function of dimension.

Now that we have introduced pure random search where each sample is independent of each other, let us turn our attention to a contrasting method, pure adaptive search, where each sample depends on the one immediately before it.

2. Pure Adaptive Search (PAS)

Pure adaptive search was first introduced by Patel, et al. [125] for convex problems. The analysis for pure adaptive search was extended to continuous global optimization problems by Zabinsky and Smith [188], and extended to finite discrete global optimization problems by Zabinsky, et al. [191].

Pure adaptive search generates a sequence of points uniformly distributed in the improving level set associated with each previous point. Let X_0 be the initial point generated within the feasible region S according to a uniform distribution ν, and let $W_0 = f(X_0)$ be the associated objective function value. The next point is generated from a truncated uniform distribution over the region of improvement, the set of points with objective function values less than W_0. The method proceeds iteratively in this fashion until a stopping criterion is satisfied.

The pure adaptive search algorithm is stated more formally below.

Pure Adaptive Search (PAS)

Step 0. Initialize $X_0 \in S$ according to a uniform distribution on S. Set $k = 0$. Set $W_0 = f(X_0.)$

Step 1. Generate X_{k+1} according to a uniform distribution on the improving set $S_k = \{x : x \in S \text{ and } f(x) < W_k\}$. Set $W_{k+1} = f(X_{k+1})$.

Step 2. If a stopping criterion is met, stop. Otherwise increment k and return to Step 1.

An important feature of the algorithm is that the iterates are consistently improving. The name *pure adaptive search* is intended to highlight the feature of improvement, because each iteration "adapts" to the current level set by restricting its search domain to improving points. In [191] where PAS was defined on a discrete finite domain, a distinction was made whether the improving set was defined with "<" or "$\leq$" in Step 1. When the improving set is defined so that the iterates are *strictly* improving,

$$S_k = \{x : x \in S \text{ and } f(x) < W_k\}$$

the version is called *strong PAS*. When the improving set includes points of equal objective function value,

$$S_k = \{x : x \in S \text{ and } f(x) \leq W_k\}$$

it is called *weak PAS*. When PAS is applied to a continuous domain, the difference between strong and weak PAS is indistinguishable, as long as the probability of sampling the set of points of equal value $\{x : x \in S \text{ and } f(x) = W_k\}$ is zero. In the rest of the book, we use strong PAS with the strict inequality. However, the results hold for weak PAS on a continuous problem if we assume that the objective function $f(x)$ has no "flat spots," i.e., for any measurable set A in the range of f with Lebesgue measure $\nu(A) = 0$, the Lebesgue measure of its preimage $\nu(f^{-1}(A))$ also must equal 0. In this situation, $f(X)$ is an absolutely continuous random variable whenever X is uniform over the domain of f. This assumption is relaxed in Chapter 3 for hesitant adaptive search.

The statement of PAS uses a uniform distribution as the underlying sampling distribution in Step 1, but this can be replaced by a general probability measure δ on S. In Chapter 3, PAS with a general probability measure δ can be seen as a special case of hesitant adaptive search, so the more general analysis is deferred to Chapter 3.

To illustrate pure adaptive search, consider Figure 2.2. The feasible region S is the box in two dimensions, and three sample points are pictured. The first point, $X_0 = A$ is sampled uniformly from S, then $X_1 = B$ is sampled from the improving set $S(W_0)$, and $X_2 = C$ is sampled from the improving set $S(W_1)$.

The feature of improvement that best characterizes PAS is also the feature that makes implementation extremely difficult. At this point in time, there is no known direct implementation of PAS, because it is very difficult to generate a point in each arbitrarily shaped improving set. If each improving set was a simple geometric figure, such as a hypercube, it could be implemented directly, but for an arbitrary set of a global optimization problem, it is extremely difficult. Approximation methods based on Hit-and-Run [157] are discussed in Chapter 6.

One way to imagine implementing PAS is to implement PRS and use a von Neumann acceptance/rejection scheme to select the improving points. For example, the initial point X_0 is generated uniformly in the domain, and the next point should be generated uniformly in the set of improving points associated with X_0. Instead, generate a point again uniformly in the domain, and reject it if it does not happen to fall in the improving region. In this way, the total number of points generated

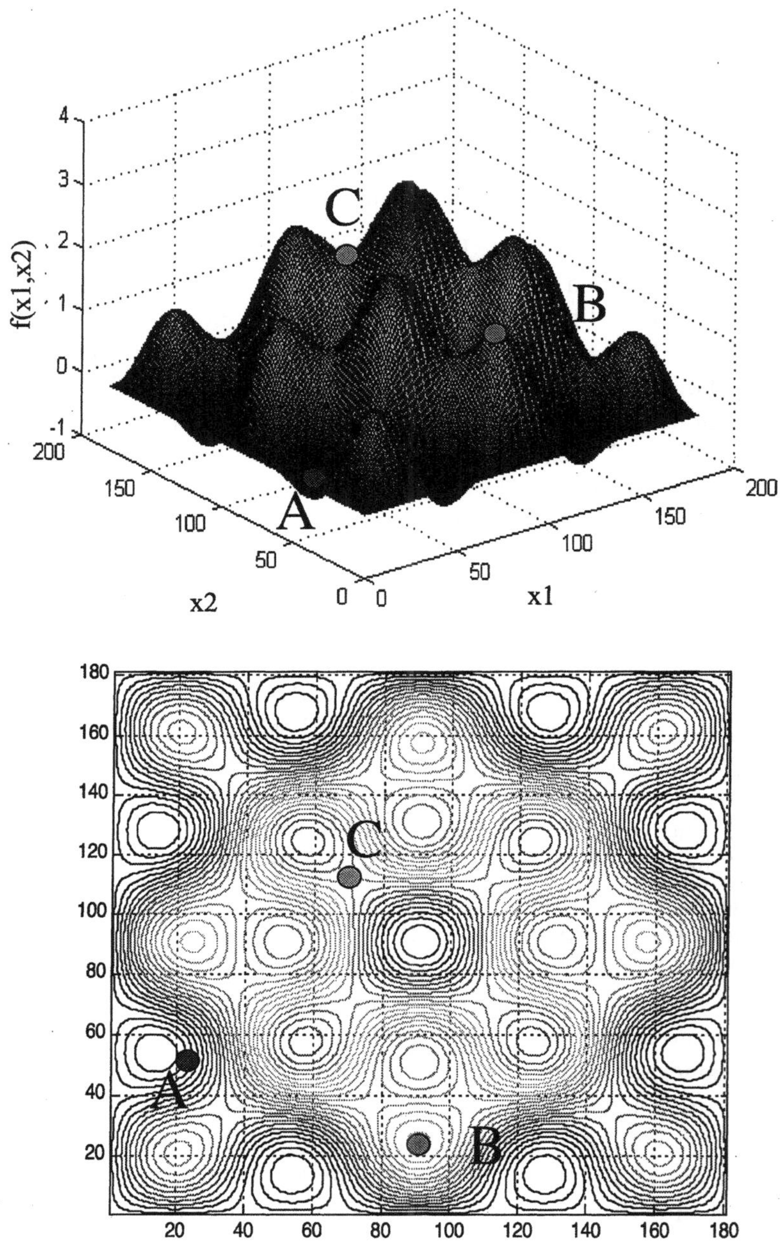

Figure 2.2. Illustration of pure adaptive search on a continuous problem.

is equivalent to PRS, and the accepted points (or records) will be a sequence of PAS. As shown in the next section, the iterates of PAS are stochastically equivalent to the record values of PRS. This also accounts for the exponential difference in complexity between pure random search and pure adaptive search.

The purpose of defining and studying PAS is to gain insight into what makes an efficient algorithm, and to spur the development of new algorithms. Later in this chapter we show that, if PAS could be implemented using a uniform distribution, the number of iterations would be linear in dimension. This suggests that the feature of consistent improvement is sufficient for an efficient algorithm.

3. Comparison of PRS and PAS

A comparison of pure adaptive search with pure random search highlights the difference between a completely "blind" search, and an adaptive "blind" search. The two methods use the same underlying uniform distribution, so the only difference is adapting the search to the region of improving points. We now establish a fundamental relationship between the iterates of PAS and PRS. The lemma below, provided originally by Zabinsky and Smith [188, Lemma 3.1] states that the *record values* of pure random search are equal in distribution to the values generated by pure adaptive search. An extension to the lemma using record values for hesitant adaptive search is mentioned in Chapter 3.

DEFINITION 2.1 *Iteration i is said to be a* record *of the sequence* $\{Y_k, k = 0, 1, \ldots\}$, *if* $Y_i < \min(Y_0, Y_1, \ldots, Y_{i-1})$, *for* $i = 1, 2, \ldots$. *The corresponding value Y_i is called a* record value.

Figure 2.3 illustrates the record values for an example of pure random search. The sequence of PRS points shown is compatible with the example used in Table 2.1 and in Figure 2.1. In Figure 2.3, the five circled points are record values. The zeroth record value is always Y_0, and in the figure Y_1, Y_3, Y_6, and Y_{10} are also record values.

The following lemma states that the record values of PRS objective function values, denoted $Y_{R(k)}$, are equal in distribution to the iterates of PAS, W_k. An intuitive understanding of Lemma 2.1 follows from the idea of using a simple acceptance-rejection approach to generate the PAS sequence $\{W_k\}$ by selecting the record values $\left\{Y_{R(k)}\right\}$ of PRS. The first two values in Figure 2.3 are $Y_0 = 6$ and $Y_1 = 4$, and those are both record values, so $Y_{R(0)} = 6$ and $Y_{R(1)} = 4$. The next point $Y_2 = 5$ is not a record value, but $Y_3 = 3$ is a record value, hence $Y_{R(2)} = 3$. Similarly the next record value occurs at $Y_6 = 2, Y_{R(3)} = 2$, and the last record value in the figure is at $Y_{10} = 1$, so $Y_{R(4)} = 1$.

LEMMA 2.1 *For the global optimization problem (P) in a continuous or discrete form, the stochastic process for PAS $\{W_k, k = 0, 1, \ldots\}$ is equal*

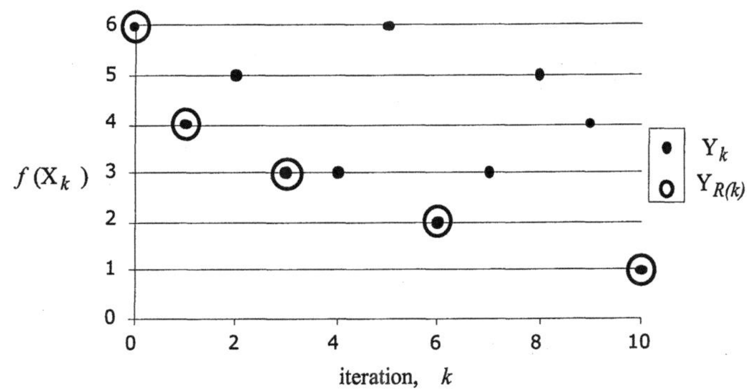

Figure 2.3. Illustration of pure adaptive search as record values of pure random search.

in distribution to the process $\left\{Y_{R(k)},\ k = 0, 1, \ldots\right\}$ *where* $R(k)$ *is the* k^{th} *record of the PRS sequence* $\{Y_k,\ k = 0, 1, \ldots\}$, *i.e.,*

$$\{W_k\} \sim \left\{Y_{R(k)}\right\}$$

and specifically,

$$P\left(W_k \leq y\right) = P\left(Y_{R(k)} \leq y\right), \qquad \text{for } k = 0, 1, \ldots \text{ and } y_* \leq y \leq y^*.$$

Proof: The proof is given in [188] and relies on the fact that the conditional probabilities are equal,

$$P\left(Y_{R(k+1)} \leq y | Y_{R(k)} = x\right) = \nu(S(y))/\nu(S(x)) = P\left(W_{k+1} \leq y | W_k = x\right)$$

for $k = 0, 1, \ldots$, and x, y such that $y_* \leq y \leq x \leq y^*$. The conditional distributions are used to show that the marginal distributions are equivalent, and induction is used to provide the result.

First, we show that the conditional distributions are equal. Let k be any iteration, $k = 0, 1, \ldots$, and if $x > y$, both conditional probabilities equal one, so consider x, y be such that $y_* \leq y \leq x \leq y^*$. Now,

$$\begin{aligned} &P\left(Y_{R(k+1)} \leq y | Y_{R(k)} = x\right) \\ &= P\left(Y_{R(k)+1} \leq y | Y_{R(k)} = x\right) \\ &\quad + P\left(Y_{R(k)+2} \leq y, Y_{R(k)+1} \geq x | Y_{R(k)} = x\right) + \cdots \end{aligned}$$

and since the Y_k are independent and $R(k)$ is a stopping time for all k,

$$= P\left(Y_{R(k)+1} \leq y\right) + P\left(Y_{R(k)+2} \leq y\right) \cdot P\left(Y_{R(k)+1} \geq x\right) + \cdots$$

and since the Y_k are identically distributed, i.e., $Y_k \sim Y_1$ for $k \geq 1$,

$$\begin{aligned} &= P(Y_1 \leq y) \sum_{i=0}^{\infty} P(Y_1 \geq x)^i \\ &= \frac{P(Y_1 \leq y)}{1 - P(Y_1 \geq x)} \\ &= \nu(S(y))/\nu(S(x)), \end{aligned}$$

which is the conditional distribution of improvement for pure adaptive search,

$$= P(W_{k+1} \leq y | W_k = x).$$

Thus, we have

$$P\left(Y_{R(k+1)} \leq y | Y_{R(k)} = x\right) = P(W_{k+1} \leq y | W_k = x).$$

Next, induction is used to show that the unconditional distributions are equal. By definition, $R(0) = 0$ and $Y_0 = W_0 = y^*$, thus $Y_{R(0)} = W_0$. Now for k=1,

$$\begin{aligned} P\left(Y_{R(1)} \leq y\right) &= P\left(Y_{R(1)} \leq y | Y_0 = y^*\right) \\ &= P(W_1 \leq y | W_0 = y^*) \\ &= P(W_1 \leq y) \end{aligned}$$

for all $y_* \leq y \leq y^*$ and hence,

$$Y_{R(1)} \sim W_1.$$

Let k be an integer greater than 1, and suppose $Y_{R(i)} \sim W_i$ for $i = 0, 1, \ldots, k$. Then,

$$\begin{aligned} P\left(Y_{R(k+1)} \leq y\right) &= E\left[P\left(Y_{R(k+1)} \leq y | Y_{R(k)}\right)\right] \\ &= \int_0^x P\left(Y_{R(k+1)} \leq y | Y_{R(k)} = x\right) dF_{Y_{R(k)}}(x) \end{aligned}$$

and using the equality of the conditional distributions and the induction hypothesis

$$\begin{aligned} &= \int_0^x P(W_{k+1} \leq y | W_k = x) dF_{W_k}(x) \\ &= E[P(W_{k+1} \leq y | W_k)] \\ &= P(W_{k+1} \leq y) \qquad \text{for all } y_* \leq y \leq y^*. \end{aligned}$$

Therefore,

$$Y_{R(k+1)} \sim W_{k+1}$$

and by induction the two sequences are equal in marginal distribution. Finally, by the equality of conditional and marginal distributions, the two sequences are equal in joint distribution. ■

As suggested in Figure 2.3, the time between record values usually increases. Early in a sequence of points, it may be easy to obtain a record value, however later in a sequence, it is less likely to obtain a record value, so the time between record values tends to increase. In fact, this observation indicates that using PRS to implement PAS is not an efficient strategy. The next theorem from [188, Theorem 3.2] emphasizes this, and proves that there is an exponential difference between the number of PRS points and the number of their record values.

THEOREM 2.2 *Let k and $R(k)$ be respectively the number of PAS and PRS iterations needed to attain an objective function value of y or better, for $y_* \leq y \leq y^*$. Then*

$$R(k) = e^{k+o(k)} \qquad \textit{with probability 1}$$

where $\lim_{k\to\infty} o(k)/k = 0$.

Proof: This proof is given in [188]. We have by definition that

$$W_{k-1} > y \geq W_k.$$

However by the previous lemma, this holds for a given k with the same probability that

$$Y_{R(k-1)} > y \geq Y_{R(k)}$$

holds. From [53, page 298], the records $R(k)$ of a sequence of continuous independent and identically distributed random variables satisfy

$$\lim_{k\to\infty} \frac{\ln R(k)}{k} = 1 \qquad \text{with probability 1.}$$

This implies that

$$\ln R(k) = k + o(k) \qquad \text{with probability 1}$$

and thus,

$$R(k) = e^{k+o(k)} \qquad \text{with probability 1.}$$ ■

The result in Theorem 2.2 states that the number of PRS iterations needed to reach the kth minimum $R(k)$ is exponentially growing in k, the number of iterations of PAS needed to reach an equivalent minimum. Thus the complexity of PRS is exponentially greater than that of PAS. Of course, each iteration of PAS may be more difficult than an iteration of PRS since the search region changes with each iteration. The simple device of forcing improvement on PRS achieves an exponential improvement in the number of iterations required.

4. Distribution of Improvement for PAS

Now we characterize the performance of pure adaptive search. The following section is devoted to a continuous problem and the subsequent section is devoted to developing analogous results for a finite discrete problem. For PAS on a continuous problem, we show that the number of PAS iterations is a nonhomogeneous Poisson process. This provides the distribution of the number of PAS iterations until a level of accuracy y is reached $N(y)$. A simple expression for the expected number of iterations and the variance is given in Corollary 2.5. The distribution of PAS objective function values $\{W_k, k = 0, 1, \ldots\}$ is provided in Corollary 2.6.

4.1 Continuous PAS Distribution

The assumptions on the global optimization problem for this section are that S is a convex, compact subset of $\mathbb{R}^n$, f is a real-valued continuous function defined over S, and $f(x)$ has no "flat spots," as discussed after PAS was defined. A more general sampling measure δ is allowed in Chapter 3.

The characterization of PAS as a nonhomogeneous Poisson process is stated in the following theorem, using a slight modification of the number of iterations. The theorem is based on relative improvement which is a transformation of the objective function value. Let Z_k be the *relative improvement* obtained on the kth iteration of PRS where $Z_k = (y^* - Y_k)/(Y_k - y_*)$, and recall that $p(y) = \nu(S(y))/\nu(S)$ is the probability of sampling the y-level set on a single iteration of pure random search. Let $M(z)$ be the number of PAS iterations achieving a relative improvement of z or less.

Figure 2.4 illustrates the correspondence between the objective function values y of PRS and the relative improvement values z, using the relationship $z = (y^* - y)/(y - y_*)$. Notice that when $y = y^*$, then $z = 0$, and when y is midway between y^* and y_*, then $z = 1$, and as y approaches y_*, then z approaches $+\infty$.

We present a result from Ross [142] that is needed in the proof of Theorem 2.4.

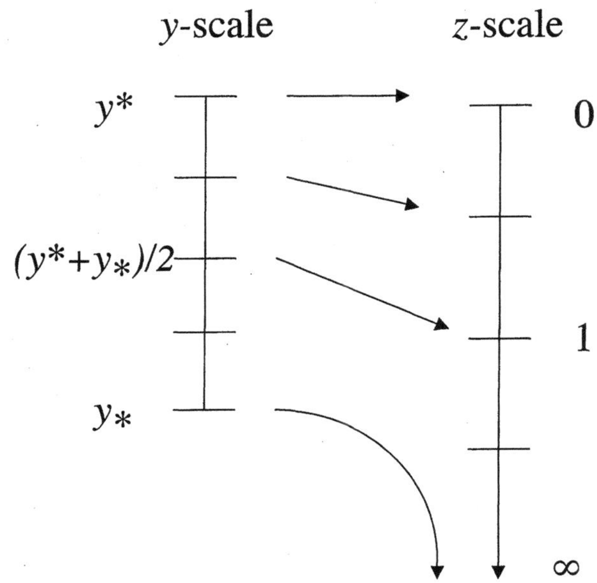

Figure 2.4. Relative improvement of $z = (y^* - y)/(y - y_*)$.

LEMMA 2.3 *Let $Z_1, Z_2, \ldots$ denote a sequence of independent and identically distributed nonnegative continuous random variables, whose hazard rate function is given by $\lambda(z)$, $\lambda(z) = f(z)/(1 - F(z))$ where f and F are respectively the density and cumulative distribution function of Z. Let $M(z)$ denote the number of record values (maximum) of $\{Z_i,\ i = 1, 2, \ldots\}$ less than or equal to z.*

Then, $\{M(z), z \geq 0\}$ is a nonhomogeneous Poisson process with intensity function $\lambda(z)$ and mean value function $m(z) = \int_0^z \lambda(s)ds$.

Proof: See [142, page 47]. ■

The following theorem, from Zabinsky and Smith [188, Theorem 4.2], establishes that PAS can be characterized by a nonhomogeneous Poisson process.

THEOREM 2.4 *Let $M(z)$ equal the number of PAS iterations achieving a relative improvement at most z for $z \geq 0$. Then $\{M(z), z \geq 0\}$ is a nonhomogeneous Poisson process with mean value function*

$$m(z) = \ln\left(1/p((y^* + zy_*)/(1 + z))\right)$$

for $0 \leq z < \infty$.

Proof: This proof follows that in [188]. The definition of relative improvement Z_k implies that $(y^* - Y_0)/(Y_0 - y_*) = 0$, and $\{Z_k,\ k = 1, 2, \ldots\}$

are independent identically distributed nonnegative continuous random variables. The cumulative distribution function F of Z_k, $k = 1, 2, \ldots$ can be written in terms of $p(y)$, as follows. For $z \geq 0$,

$$\begin{aligned} F(z) &= P(Z_k \leq z) \\ &= P(Y_k \geq (y^* + zy_*)/(1+z)) \\ &= \begin{cases} 0 & \text{if } z < 0 \\ 1 - p((y^* + zy_*)/(1+z)) & \text{if } 0 \leq z < \infty. \end{cases} \end{aligned}$$

$M(z)$ counts the number of records of $\{Z_k, \ k = 0, 1, \ldots\}$ with values less than or equal to z. By Lemma 2.1, $M(z)$ is equal in distribution to the number of PAS iterations achieving a relative improvement of z or less. Invoking Lemma 2.3, $\{M(z), z \geq 0\}$ is a nonhomogeneous Poisson process with intensity function $\lambda(z) = f(z)/(1 - F(z))$, where $f(z)$ is the density function associated with the cumulative distribution function $F(z)$, and mean value function $m(z) = \int_0^z \lambda(s)ds$.

To express the mean value function in terms of $p(y)$, by definition

$$m(z) = \int_0^z \lambda(s)ds$$

where $\lambda(s) = f(s)/(1 - F(s))$. Making the substitution $t = 1 - F(s)$ yields

$$\begin{aligned} m(z) &= \int_1^{1-F(z)} -(1/t)dt \\ &= -\ln t|_1^{1-F(z)} \\ &= -\ln(1 - F(z)) \end{aligned}$$

for $z \geq 0$. Now,

$$1 - F(z) = p((y^* + zy_*)/(1+z))$$

and hence,

$$m(z) = \ln(1/p((y^* + zy_*)/(1+z)))$$

for $0 \leq z < \infty$. ■

Instead of using relative improvement, a more natural measure of the performance of pure adaptive search is the number of iterations $N(y)$ required to achieve an objective function value of y or better. Since an objective function value of y corresponds to a relative improvement of $z = (y^* - y)/(y - y_*)$, it follows that

$$N(y) = M((y^* - y)/(y - y_*)) + 1.$$

Since $M(z)$ counts the iterations less than or equal to z, and $N(y)$ counts the number of iterations needed to achieve at least y, $N(y)$ is one more than $M(z)$. For example, suppose Y_0 and Y_1 have values greater than y and Y_2 has a value less than y, then $N(y) = 3$ because the third point is below the value of y. The corresponding Z_0 and Z_1 have values less than z, and Z_2 is greater than z, so the number of iterations achieving a relative improvement of z or less is only 2, $M(z) = 2$.

The distribution of $N(y)$ and expressions for the expected value and variance are stated in the following corollary (first appearing in [188, Corollary 5.1]).

COROLLARY 2.5 *The cumulative distribution function of the number of iterations of PAS needed to achieve a value of y or better, $N(y)$, is given by*

$$P(N(y) \leq k) = \sum_{i=0}^{k-1} \frac{p(y)(\ln(1/p(y)))^i}{i!} \tag{2.3}$$

for $k = 1, 2, \ldots$ and $y_* \leq y \leq y^*$, and

i) the expected value of $N(y)$ is

$$E[N(y)] = 1 + \ln(1/p(y)),$$

ii) the variance of $N(y)$ is

$$Var(N(y)) = \ln(1/p(y)).$$

Proof: This proof is given in [188]. The cumulative distribution follows immediately from the relationship,

$$N(y) = M(z) + 1$$

where $z = (y^* - y)/(y - y_*)$, for all $y^* \leq y \leq y_*$. Because $M(z)$ is a Poisson distributed random variable with mean $m(z)$ and variance $m(z)$, we have

$$P(N(y) \leq k) = P(M(z) \leq k-1) = \sum_{i=0}^{k-1} \frac{e^{-m(z)} m(z)^i}{i!}$$

with

$$E[N(y)] = E[M(z)] + 1 = 1 + m(z),$$
$$\mathrm{Var}(N(y)) = \mathrm{Var}(M(z)) = m(z)$$

where $m(z) = \ln(1/p(y))$. ■

As seen in Theorem 2.4 and Corollary 2.5, performance measures of PAS depend on the function $p(y)$ for $y_* \leq y \leq y^*$, where $p(y)$ is the probability of obtaining an objective function value between y and y_* when selecting a feasible point at random according to a uniform distribution. In Section 5, we derive a bound on $p(y)$ for the class of global optimization problems with objective functions that satisfy the Lipschitz condition over a convex feasible region. The bound is a function of the dimension n of the problem; the Lipschitz constant K_f of the objective function; and the maximum diameter D_S of the feasible region, where $D_S = \max\{\|w - v\|, w, v \in S\}$.

It is now an easy matter to obtain the distribution of the objective function values obtained through pure adaptive search. In particular, since

$$W_k \leq y \text{ if and only if } N(y) \leq k$$

we can use Corollary 2.4 to obtain the distribution of the objective function values W_k. The following corollary is from [188, Theorem 4.3].

COROLLARY 2.6

$$P(W_k \leq y) = \sum_{i=0}^{k-1} \frac{p(y)\left(\ln\left(1/p(y)\right)\right)^i}{i!}$$

for $k = 1, 2, \ldots$ *and* $y_* \leq y \leq y^*$.

Proof: To justify that $W_k \leq y$ if and only if $N(y) \leq k$, first suppose that $W_k \leq y$. Then the kth point fell in $S(y)$, but it may not have been the first one, so $N(y) \leq k$. Supposing $N(y) \leq k$, then the first point landing in $S(y)$ occurred at iteration k or earlier, so W_k must be in $S(y)$. Then, for $k = 1, 2, \ldots$ and $y_* \leq y \leq y^*$,

$$\begin{aligned} P(W_k \leq y) &= P(N(y) \leq k) \\ &= \sum_{i=0}^{k-1} p(y)\Big(\ln\left(1/p(y)\right)\Big)^i / i! \end{aligned}$$

by Corollary 2.5. ∎

There are several problem classes in the literature where the asymptotic distribution for large sample sizes of PRS and PAS have been obtained (see for example [7, 35, 39, 124, 125]). The result in Corollary 2.6 is particularly striking in that it provides the exact distribution of values generated by PAS for all sample sizes and all continuous global optimization problems.

4.2 Finite PAS Distribution

The PAS algorithm has also been analyzed for a finite global optimization problem [191], with analogous results. This section develops analytical expressions for the expected number of PAS iterations, as well as the distribution of the objective function value.

The finite discrete problem of this section assumes there are M distinct objective function values, denoted $y_1 < y_2 < \cdots < y_M$. Notice that there may be more than M points in S. For $k = 0, 1, \ldots$, let the random variable Y_k be the objective function value on the kth iteration of PRS. Note that $Y_0, Y_1, \ldots$ are independent and identically distributed. Given a probability measure δ on S, typically a uniform distribution, we define its related probability measure $\pi = (\pi_1, \ldots, \pi_M)$ on the range of f as follows. Let π_j be the probability that any iteration of pure random search attains a value of y_j, that is $\pi_j = P(Y_0 = y_j) = \delta(f^{-1}(y_j))$ for $j = 1, 2, \ldots, M$. Also, let $p_j = \sum_{i=1}^{j} \pi_i$ be the probability that PRS attains a value of y_j or less.

As in the continuous case, let the random variable W_k be the objective function value on the kth iteration of PAS, and let $R(k)$ be the epoch of the kth record of PRS. Then as in the continuous problem, PAS consists of the record values of pure random search. That is, W_k is stochastically equivalent to $Y_{R(k)}$. This is directly analogous to Lemma 2.1 and is proved identically.

The distribution of pure adaptive search on a finite domain depends on whether the improving set, in Step 1 of the algorithm, is defined so that the iterates are *strictly* improving, which is called *strong PAS*, or the set includes points of equal value, which is called *weak PAS*. The stochastic process $\{W_k, k = 0, 1, \ldots\}$ for either strong or weak PAS can be modeled as a Markov chain with states $y_1, \ldots, y_M$, where state y_1 represents the global optimum. The initial probability distribution for W_0 is given by π. In standard Markov chain terminology [82], y_1 is the absorbing state of this chain and all other states are transient. Finite PAS converges when the chain reaches the absorbing state. The expected number of iterations to convergence can be expressed in terms of the transition matrix of the Markov chain, and represent the time until first sampling the global optimum. Notice this does not include any stopping rule.

The M by M one-step transition matrix P, in standard form, with element (i, j) the one-step probability $P(W_{k+1} = y_j | W_k = y_i)$ is given below, first for strong PAS, and then for weak PAS [191]. Notice that the matrices are similar and strong PAS is strictly improving, while weak PAS may retain an equal objective function value.

For strong PAS, the one-step transition matrix P is

$$\begin{bmatrix} 1 & 0 & 0 & \dots & 0 & 0 \\ \pi_1/p_1 & 0 & 0 & \dots & 0 & 0 \\ \pi_1/p_2 & \pi_2/p_2 & 0 & \dots & 0 & 0 \\ \vdots & \vdots & \vdots & & \vdots & \vdots \\ \pi_1/p_{M-1} & \pi_2/p_{M-1} & \pi_3/p_{M-1} & \dots & \pi_{M-1}/p_{M-1} & 0 \end{bmatrix}$$

and for weak PAS, the one-step transition matrix P is

$$\begin{bmatrix} 1 & 0 & 0 & \dots & 0 & 0 \\ \pi_1/p_2 & \pi_2/p_2 & 0 & \dots & 0 & 0 \\ \pi_1/p_3 & \pi_2/p_3 & \pi_3/p_3 & \dots & 0 & 0 \\ \vdots & \vdots & \vdots & & \vdots & \vdots \\ \pi_1/p_M & \pi_2/p_M & \pi_3/p_M & \dots & \pi_{M-1}/p_M & \pi_M/p_M \end{bmatrix}$$

To highlight the difference, consider $P(W_{k+1} = y_2 | W_k = y_3)$. For strong PAS, $P(W_{k+1} = y_2 | W_k = y_3) = \pi_2/p_2$ because W_{k+1} must be either y_1 or y_2 to be strictly improving and $p_2 = \pi_1 + \pi_2$. For weak PAS, $P(W_{k+1} = y_2 | W_k = y_3) = \pi_2/p_3$ because there is a possibility that W_{k+1} remains at y_3, reflected by the divisor $p_3 = \pi_1 + \pi_2 + \pi_3$.

It follows that the distribution of sampling objective function value y_i on the kth iteration, $P(W_k = y_i)$, can be obtained by calculating the k-step transition probabilities using the appropriate P matrix, thus providing an analogous result to Corollary 2.6. This result was observed in [191] and is summarized below.

COROLLARY 2.7 *For strong or weak PAS on a finite global optimization problem, with $k = 1, 2, \dots$ and $i = 1, 2, \dots, M$*

$$P(W_k = y_i) = i^{\text{th}} \textit{entry of } \pi P^k,$$

with the appropriate P matrix.

The expression for the expected number of iterations $E(N(y_*))$ to solve the finite global optimization problem is from [191, Theorem 1].

THEOREM 2.8 *The expected number of iterations to solve the finite optimization problem is*

i) $E[N(y_*)] = 1 + \sum_{j=2}^{M} \pi_j/p_j$ *for strong PAS and*
ii) $E[N(y_*)] = 1 + \sum_{j=2}^{M} \pi_j/p_{j-1}$ *for weak PAS*

where $p_j = \sum_{i=1}^{j} \pi_i$ and $y_* = y_1$.

Proof: A simple direct probabilistic argument is reproduced here from [191], although the Markov transition matrix can also be used to prove the result.

First, express the number of iterations required to solve the finite optimization problem, as $N(y_*) = 1 + X_2 + \cdots + X_M$, where X_j is the number of iterations spent in state y_j. Thus

$$\begin{aligned} E[N(y_*)] &= 1 + E[X_2] + \cdots + E[X_M] \\ &= 1 + E[X_2|V_2]P(V_2) + \cdots + E[X_M|V_M]P(V_M) \end{aligned}$$

where V_j is the event that state y_j is visited. Now

$$\begin{aligned} P(V_j) &= P(V_j \cap \{W_0 > y_j\}) + P(V_j \cap \{W_0 = y_j\}) \\ &= \sum_{i=j+1}^{M} P(V_j \cap B_{i,j} \cap \{W_0 > y_j\}) + P(V_j \cap \{W_0 = y_j\}) \end{aligned}$$

where $B_{i,j}$ is the event that state y_i is visited immediately before a state less than or equal to y_j is visited. Since $P(V_j|B_{i,j} \cap \{W_0 > y_j\}) = \pi_j/p_j$ and $P(W_0 = y_j) = \pi_j$, we have

$$\begin{aligned} P(V_j) &= \sum_{i=j+1}^{M} P(V_j|B_{i,j} \cap \{W_0 > y_j\})P(B_{i,j}|W_0 > y_j)P(W_0 > y_j) \\ &\quad + P(V_j|W_0 = y_j)P(W_0 = y_j) \\ &= (1 - p_j)\frac{\pi_j}{p_j} \sum_{i=j+1}^{M} P(B_{i,j}|W_0 > y_j) + \pi_j \\ &= (1 - p_j)\frac{\pi_j}{p_j} + \pi_j = \frac{\pi_j}{p_j}. \end{aligned}$$

For strong PAS, $E[X_j|V_j] = 1$, and $E[N(y_*)] = 1 + \sum_{j=2}^{M} \pi_j/p_j$ as required. For weak PAS, once state y_j is left, it is never re-entered. As with a geometric distribution, $E[X_j|V_j]$ then equals the inverse of the probability of leaving state y_j on a given iteration,

$$E[X_j|V_j] = (p_{j-1}/p_j)^{-1} = p_j/p_{j-1}.$$

This yields $E[N(y_*)] = 1 + \sum_{j=2}^{M} \pi_j/p_{j-1}$ as required. ■

The exact expressions given in Theorem 2.8 are valid for an arbitrary sampling distribution reflected by π and p. We are able to describe the number of iterations $N(y)$ and the objective function values W_k for strong and weak PAS in the finite case through the underlying sampling distribution p_j, which is analogous to $p(y)$ for PAS in the continuous case. In the next section, we obtain bounds on $p(y)$ and p_j to further characterize performance of PAS.

5. Linearity Result for PAS

In this section, we first derive a bound on $p(y)$ for the class of continuous global optimization problems with objective functions that satisfy the Lipschitz condition over a convex feasible region. This bound on $p(y)$ is then used to obtain a bound on the expected number of iterations, and show that the expected number of iterations is *linear* in the dimension of the problem.

An analogous bound on the expected number of iterations of PAS for the finite discrete domain is also developed. In order to compare the expressions, the analysis is performed on an n-dimensional lattice $\{1, \ldots, K\}^n$. The expected number of iterations on this finite problem is also shown to be *linear* in dimension.

A bound on $p(y)$ for a continuous problem, first given in [188, Lemma 5.2], follows.

THEOREM 2.9 *For a global optimization problem (P) over a convex feasible region S in n dimensions with diameter D and Lipschitz constant K for objective function f,*

$$p(y) \geq \left((y - y_*) / KD \right)^n \qquad \textit{for } y_* \leq y \leq y^*.$$

Proof: The proof is reproduced from [188]. To obtain a lower bound on $p(y)$, we construct two intermediate bounds in Equations 2.4 and 2.7. Let

$$\begin{aligned}
r &= (y^* - y_*)/K \\
B_r &= n\text{-dimensional ball centered at } x_* \text{ with radius } r \\
B_D &= n\text{-dimensional ball centered at } x_* \text{ with radius } D \\
g(x) &= \begin{cases} K\|x - x_*\| + y_* & \text{if } x \in B_r \\ y^* & \text{otherwise} \end{cases} \\
&= \begin{cases} \inf\,[y : (x,y) \in \text{convex hull of } (x_*, y_*) \text{ and } (B_r, y^*)] & \text{if } x \in B_r \\ y^* & \text{otherwise} \end{cases} \\
G(y) &= \{x : x \in S \textit{ and } g(x) \leq y\} \\
S_r &= B_r \cap S \\
h(x) &= \begin{cases} \inf\,[y : (x,y) \in \text{convex hull of } (x_*, y_*) \text{ and } (S_r, y^*)] & \text{if } x \in S_r \\ y^* & \text{otherwise} \end{cases} \\
H(y) &= \{x : x \in S \textit{ and } h(x) \leq y\}.
\end{aligned}$$

For a one-dimensional geometrical interpretation of the functions $f(x)$, $g(x)$, $h(x)$ and sets B_r, S_r, S, refer to Figure 2.5. Notice that two expressions are given for $g(x)$. These expressions are equivalent in defining $g(x)$, which can be easily verified by comparing the ratio $\|x - x_*\|/r$ to the ratio $y/(y^* - y_*)$. The sets $G(y)$ and $H(y)$ are the level sets of the functions $g(x)$ and $h(x)$ respectively.

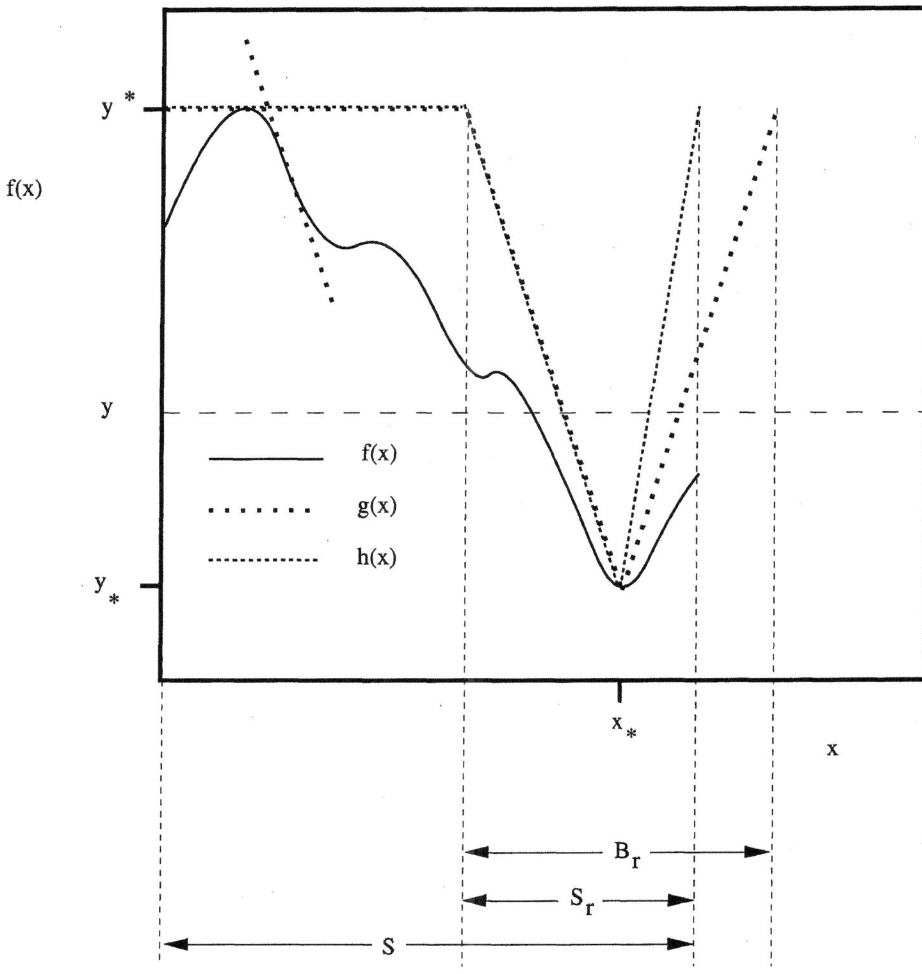

Figure 2.5. A one-dimensional problem to illustrate the bound on $p(y)$.

We first show that the level sets are nested,

$$H(y) \subseteq G(y) \subseteq S(y)$$

for any $y_* \leq y \leq y^*$. We begin by considering any $x \in H(y)$ for any $y_* \leq y \leq y^*$. By definition of $H(y)$, $x \in S$ and $h(x) \leq y$. There are two cases, either $x \in S_r$ or $x \notin S_r$.

If $x \in S_r$, then

$$h(x) = \inf\left[y : (x, y) \in \text{convex hull of } (x_*, y_*) \text{ and } (S_r, y^*)\right].$$

Since $S_r = B_r \bigcap S$, $x \in B_r$, so

$$g(x) = \inf\left[y : (x, y) \in \text{convex hull of } (x_*, y_*) \text{ and } (B_r, y^*)\right].$$

Also, $S_r \subseteq B_r$ implies that $h(x) \geq g(x)$, thus $x \in G(y)$. Now, because $f(x)$ satisfies the Lipschitz condition, we have

$$|f(x) - f(x_*)| \leq K\|x - x_*\|$$

for any $x \in S$, or equivalently

$$f(x) \leq K\|x - x_*\| + y_*$$

which equals $g(x)$ using the first expression for defining $g(x)$. Thus $f(x) \leq g(x)$ and $x \in S(y)$.

If $x \notin S_r$, then

$$h(x) = y^*.$$

Also, because $x \in S$, we have $x \notin B_r$, and thus

$$g(x) = y^*.$$

And $f(x) \leq y^*$, so again, $x \in H(y)$, $x \in G(y)$, and $x \in S(y)$. Combining both cases gives us

$$H(y) \subseteq G(y) \subseteq S(y).$$

Now, using this result yields,

$$\begin{aligned} p(y) &= \nu(S(y))/\nu(S) \\ &\geq \nu(G(y))/\nu(S) \\ &\geq \nu(H(y))/\nu(S) \\ &= (\nu(H(y))/\nu(S_r))\,(\nu(S_r)/\nu(S)) \end{aligned}$$

and due to the similarity of the convex conical sets,

$$= ((y - y_*) \,/\, (y^* - y_*))^n \,(\nu(S_r)/\nu(S))\,.$$

Thus the first intermediate bound is:

$$p(y) \geq ((y - y_*) \,/\, (y^* - y_*))^n \,(\nu(S_r)/\nu(S))\,. \tag{2.4}$$

It remains to develop a lower bound on $\nu(S_r)/\nu(S)$. Before we continue, notice that for some mathematical programs, the set S_r equals the set S, and thus the bound simplifies to

$$p(y) \geq ((y - y_*) \,/\, (y^* - y_*))^n\,.$$

For example, let $f(x) = |x|$, with $S = [-2, 2]$. In this simple example, $K = 1$, $y^* - y_* = 2$, and $r = 2$. Thus we have an example with $B_r = S = S_r$, and the bound is tight. In fact, the bound in Equation 2.4

is tight for the class of convex programs, even though S_r is not necessarily equal to S for all convex programs. A proof is given in Patel, et al. [125].

We now show that $\nu(S_r)/\nu(S) \geq \nu(B_r)/\nu(B_D)$. In order to continue the proof, we need to introduce a similarity transformation. Let $\lambda(x) : \mathbb{R}^n \rightarrow \mathbb{R}^n$ be the affine function defined by

$$\lambda(x) = x_* + c\,(x - x_*)$$

for any $x \in S$ with $c = \dfrac{r}{D} = \dfrac{(y^* - y_*)}{KD}$. The transformation takes a point x in S and moves it towards x_* by a factor of c. Also, let

$$\widetilde{B}_D = \{\widetilde{x} : \widetilde{x} = \lambda(x), x \in B_D\}$$

and similarly,

$$\widetilde{S} = \{\widetilde{x} : \widetilde{x} = \lambda(x), x \in S\}.$$

The first step is to establish that

$$\widetilde{B}_D = B_r. \tag{2.5}$$

This follows directly from applying the similarity transformation λ to B_D, which shrinks a ball of radius D centered at x_* by a factor of $c = r/D$. This yields a ball of radius r centered at x_*, or B_r.

The second step is to establish that

$$\widetilde{S} \subseteq S_r. \tag{2.6}$$

To prove this, consider any $\widetilde{x} \in \widetilde{S}$. We show that $\widetilde{x} \in S$, and $\widetilde{x} \in B_r$. Now, $\widetilde{x} \in S$ because

$$\begin{aligned} \widetilde{x} &= \lambda(x) \qquad \text{for some } x \in S \\ &= x_* + c\,(x - x_*) \\ &= cx + (1 - c)x_* \end{aligned}$$

and since $x, x_* \in S$ and $0 \leq c \leq 1$, and S is convex, we conclude

$$\widetilde{x} \in S.$$

Also, $\widetilde{x} \in B_r$ because

$$\begin{aligned} \|\widetilde{x} - x_*\| &= \|\lambda(x) - x_*\| \qquad \text{for some } x \in S \\ &= \|x_* + c\,(x - x_*) - x_*\| \\ &= \|c\,(x - x_*)\,\| \\ &= c\|\,(x - x_*)\,\| \end{aligned}$$

and by the definition of $c = r/D$

$$= (r/D)\|(x - x_*)\|$$

and because $x \in S$ and D is the diameter of S,

$$\begin{aligned} &\leq (r/D)D \\ &= r, \end{aligned}$$

hence,

$$\widetilde{x} \in B_r.$$

Therefore, $\widetilde{x} \in S_r$, since $S_r = S \bigcap B_r$.

Now, using Equation 2.6 yields,

$$\begin{aligned} \nu(S_r)/\nu(S) &\geq \nu(\widetilde{S})/\nu(S) \\ &= \nu(\widetilde{B}_D)/\nu(B_D) \end{aligned}$$

because the ratio of the contents of sets is preserved under the similarity transformation, and

$$= \nu(B_r)/\nu(B_D)$$

from Equation 2.5. Therefore,

$$\nu(S_r)/\nu(S) \geq \nu(B_r)/\nu(B_D).$$

Substituting the above inequality into Equation 2.4 yields the second intermediate bound:

$$p(y) \geq ((y - y_*) / (y^* - y_*))^n (\nu(B_r)/\nu(B_D)). \tag{2.7}$$

But the ratio of the volumes of two n-dimensional balls of radii a and b is $(a/b)^n$ [83], thus

$$\begin{aligned} \nu(B_r)/\nu(B_D) &= (r/D)^n \\ &= ((y^* - y_*) / KD)^n. \end{aligned}$$

Hence, the final bound on $p(y)$ is:

$$p(y) \geq ((y - y_*) / KD)^n \tag{2.8}$$

and the proof is concluded. ■

Theorem 2.9 together with Corollary 2.5 provides the linearity result for PAS [188, Theorem 5.3].

COROLLARY 2.10 *For all global optimization problems (P) over a convex feasible region in n dimensions with diameter at most D, and with Lipschitz constant at most K,*

$$E[N(y)] \leq 1 + n[\ln(KD/(y - y_*))]$$

and

$$Var(N(y)) \leq n[\ln(KD/(y - y_*))]$$

for $y_* \leq y \leq y^*$.

Proof: As in [188], this follows immediately from Theorem 2.9, where

$$p(y) \geq ((y - y_*)/KD)^n$$

implies that

$$1/p(y) \leq (KD/(y - y_*))^n$$

and from Corollary 2.5,

$$\begin{aligned} E[N(y)] &= 1 + \ln(1/p(y)) \\ &\leq 1 + n[\ln(KD/(y - y_*))], \end{aligned}$$

and

$$\begin{aligned} \mathrm{Var}(N(y)) &= \ln(1/p(y)) \\ &\leq n[\ln(KD/(y - y_*))]. \end{aligned}$$

∎

From Corollary 2.10, we conclude that the expected number of PAS iterations grows linearly in dimension for a class of problems with finite Lipschitz constant K and feasible region diameter D. This is in contrast to PRS where from Theorem 2.2 we know that the expected number of iterations will be an exponential function of dimension n. The logarithmic term $KD/(y - y_*)$ provides the impact on performance in terms of parameters of the problem, K, D, and accuracy $y - y_*$. An exponential increase in the Lipschitz constant or the diameter gives rise to a linear increase in the corresponding number of iterations of PAS required to achieve the same value error. The effect of achieving an ϵ improvement ($\epsilon = y - y_*$) on complexity is logarithmic, $\ln(1/\epsilon)$.

The theorem states that the variance for the expected number of PAS iterations is also bounded by a linear function in dimension. This should encourage the search for practical PAS algorithms, since the coefficient of variation of the number of iterations required approaches zero as n gets large. In particular, the *actual* number of iterations required to yield a solution close to optimal is highly likely, for large dimensional problems, to be the *expected* number of iterations given in Corollary 2.10, i.e., linear in the dimension of the problem.

The linearity result of Corollary 2.10 indicates that if PAS could be implemented it would be a very good algorithm. It is interesting to note that several researchers have reported empirical linear behavior of a variety of random search algorithms including Schrack and Borowski [150], Schumer and Steiglitz [152], and Solis and Wets [159]. The test function commonly used in these three papers was a simple quadratic function, and a justification for the linear behavior was maintaining a constant probability of successful improvement. Another way to rationalize the behavior is to say that these sequential random search algorithms, on a quadratic function, are approximating the behavior of PAS.

This interpretation of approximating PAS leads to the idea of approximating uniformly distributed points in the improving level sets, which may be arbitrary nonconvex sets. Although the problem of generating a *single* point uniformly distributed in a bounded region is still unsolved (though see [28] for a quantum-computational partial solution), the problem of generating *many* points nearly uniformly distributed in a general region has met with some success [19, 24, 143, 157].

An approach to designing an efficient algorithm is to generate points that in value approximate or stochastically dominate the uniform distribution. That is, if an algorithm generates random points (uniform or not) with associated objective function values stochastically better than those of uniformly distributed points, then the same linear bound on performance will apply. Possible implementations that approximate PAS are discussed in Chapter 6.

Bounds are also developed for the finite discrete problem. An upper bound on the expected number of PAS iterations to solve the finite optimization problem can be stated in terms of π_1, the probability of sampling the global optimum with pure random search [191, Corollaries 2 and 3]. For weak PAS, the maximum hazard rate is also involved in the bound.

COROLLARY 2.11 *The expected number of iterations to solve the finite optimization problem is bounded above by,*

i) $E[N(y_*)] \leq 1 + \ln(1/\pi_1)$ *for strong PAS, and*
ii) $E[N(y_*)] \leq 1 + (M_{\text{hazard}} + 1)\ln(1/\pi_1)$ *for weak PAS*

where $M_{\text{hazard}} = \max_{j=2,\ldots,M} \pi_j / p_{j-1}$.

Proof: The proof first appears in [191]. For $0 < x < 1$, $x < -\ln(1-x)$, so for $j = 2, \ldots, M$

$$\frac{\pi_j}{p_j} < -\ln\left(1 - \frac{\pi_j}{p_j}\right) = \ln\left(\frac{p_j}{p_{j-1}}\right).$$

Therefore, from Theorem 2.8(i) for strong PAS, the expected number of iterations is less than $1 + \ln(p_2/p_1) + \ln(p_3/p_2) + \cdots + \ln(p_M/p_{M-1}) = 1 + \ln(p_M/p_1) = 1 + \ln(1/\pi_1)$.

For weak PAS the maximum hazard rate is involved in the bound. Note the term π_j/p_{j-1} appearing in Theorem 2.8(ii) can be written as $(1 + \pi_j/p_{j-1})\pi_j/p_j$ so a similar argument applies. ■

It is interesting to compare the expression for strong PAS on a finite problem, given in Corollary 2.11,

$$E[N(y_*)] \leq 1 + \ln(1/\pi_1)$$

with the expression for PAS on a continuous problem, given in Corollary 2.5,

$$E[N(y)] = 1 + \ln(1/p(y)).$$

In both cases, the reciprocal of the probability of sampling the global optimum characterizes performance within the logarithm. This expression is modified for weak PAS with another term that captures the probability of staying at exactly the same objective function value. Strong PAS has complexity of no more than M, the size of the range, because forced improvement ensures it never requires more than M iterations. This cannot be said for weak PAS (which is closer to practical algorithms). Weak PAS may require a large number of iterations when π_{i+1}/π_i is large for some i in the problem. Additional bounds for weak PAS are given in [191] for specific conditions on the ratios π_{i+1}/π_i. A generalization of PAS that includes "hesitation," or remaining at the same objective function value as in weak PAS, is analyzed in Chapter 3, and a similar expression for the expected number of iterations is derived.

In order to extend the linearity result of pure adaptive search on a continuous problem to a finite problem, we need to specify what dimension means in the context of a finite discrete problem. For this purpose, consider the special case where the domain S consists of the vertices of an n-dimensional lattice, $\{1, \ldots, K\}^n$, with distinct objective function values at all points. Then the number of distinct objective function values equals the number of points in the lattice, $M = K^n$. Assume the underlying distribution is uniform on the domain, and because each point in the domain has a unique objective function value, the distribution is also uniform on the range, $\pi_j = 1/M = 1/K^n$ for all j, $j = 1, \ldots, M$.

The following corollary, first appearing in [191, Corollary 6], provides an expression for PAS on a finite problem assuming a uniform distribution, and includes an analogous linearity result for PAS on a lattice problem.

COROLLARY 2.12 *The expected number of iterations for PAS on a finite global optimization with M distinct objective function values, and a uniform distribution on the objective function values, is*

$$\begin{array}{llcll} i) & E[N(y_*)] & = & \sum_{j=1}^{M} 1/j & \leq 1+\ln(M) & \textit{for strong PAS} \\ ii) & E[N(y_*)] & = & 1+\sum_{j=1}^{M-1} 1/j & \leq 2+\ln(M-1) & \textit{for weak PAS.} \end{array}$$

For an n-dimensional lattice, $\{1,\ldots,K\}^n$, with distinct objective function values at all points, and a uniform distribution, the expected number of iterations for strong PAS or weak PAS is bounded by

$$E[N(y_*)] \leq 2 + n(\ln(K)).$$

Proof: Reproduced from [191].
i) For strong PAS, with $\pi_j = 1/M$, we have

$$\begin{aligned} E[N(y_*)] &= 1+\sum_{j=2}^{M} \pi_j/p_j \\ &= 1+\sum_{j=2}^{M} 1/j \\ &= \sum_{j=1}^{M} 1/j \\ &\leq 1+\ln(M). \end{aligned}$$

ii) For weak PAS, with $\pi_j = 1/M$, we have

$$\begin{aligned} E[N(y_*)] &= 1+\sum_{j=2}^{M} \pi_j/p_{j-1} \\ &= 1+\sum_{j=2}^{M} 1/(j-1) \\ &= 1+\sum_{j=1}^{M-1} 1/j \\ &\leq 2+\ln(M-1). \end{aligned}$$

For either strong PAS or weak PAS on the lattice problem, with $\pi_j = 1/M$,

$$E[N(y_*)] \leq 2 + \ln(M)$$

and because $M = K^n$,

$$\leq 2 + n[\ln(K)]$$

which provides the analogous linearity result. ■

Hence, for an n-dimensional lattice, the expected number of iterations for strong or weak PAS given a uniform distribution is of order n, consistent with the linearity result for PAS on a continuous optimization problem. These performance bounds and linearity results suggest that the blind improvement of PAS on a continuous or discrete global optimization problem is extremely efficient. It has inspired the development of several random search methods (see Chapter 6), as well as extensions of the analysis to more general analyses (Chapters 3, 4, and 5).

6. Summary

The number of iterations to achieve a level of accuracy is used to compare pure random search with pure adaptive search, and it is shown that pure adaptive search is an exponential improvement on pure random search. The main result in this chapter is that the expected number of iterations of pure adaptive search is *linear* in dimension. For continuous global optimization problems satisfying a Lipschitz condition, a bound on the expected number of PAS iterations is shown to be a linear function of dimension. An analogous statement is developed for pure adaptive search applied to an n-dimensional lattice with distinct objective function values. In addition, the distribution of the iterates for pure adaptive search, on a continuous domain, is shown to be described by a nonhomogeneous Poisson process. The distribution of the iterates in the finite case is also established, using Markov chain theory. This provides a more detailed understanding of the behavior of the algorithm. The linear performance of pure adaptive search suggests that steady improvement on the objective function value, even if it is not clever and simply uniformly distributed, is extremely valuable for an algorithm.

Chapter 3

HESITANT ADAPTIVE SEARCH

Hesitant adaptive search (HAS) extends the ideas of PAS by accommodating hesitation, or pausing, at objective function values for both continuous and discrete problems. At each iteration the HAS algorithm can either generate a point in the improving region with a certain "bettering" probability, or it "hesitates" and stays at the current point. PAS is a special case of HAS, occurring when the bettering probability is equal to one. The HAS algorithm goes a step further than PAS in modeling stochastic algorithms, because it allows hesitation at each iterate which may represent rejecting a sampled point. The bettering probability depends on the objective function value, so it can capture the realistic effect of it being harder to find an improving point as the objective function value gets smaller.

Hesitant adaptive search was first introduced by Bulger and Wood [29], and an expression for the expected number of HAS iterations required to reach a specified level of accuracy was presented and derived using a first principles analysis. This expression for the expected number of HAS iterations until termination is stated in Corollary 3.7, along with the variance. The complete distribution for the number of HAS iterations for a general mixed continuous-discrete global optimization problem was developed by Wood, et al. [181], and is stated in Theorem 3.6. The analysis derived the distribution of the number of iterations of hesitant adaptive search to termination using the power of "marked Poisson processes." The material in this chapter is largely based on [29] and [181].

Pure adaptive search is a special case of HAS, and the distributions reflect this. The distribution for PAS was seen in Theorem 2.4 to be a non-homogenous Poisson process, and the distribution of HAS is linked to standard counting process theory. Intuitively, HAS can be thought of as a Poisson process for the values of the records, and the process is marked by a geometric distribution for the time spent hesitating.

1. Hesitant Adaptive Search (HAS)

The global optimization problem (P) in this chapter is taken to be very general, where S is assumed to be a measurable space, and $f : S \rightarrow \mathbb{R}$ is a measurable function. The number of iterations to first sample an objective function value of y or less is again the primary measure of performance.

Hesitant adaptive search is governed by two entities, the probability measure δ of the underlying sampling distribution on the domain S, and the bettering probability $b(y)$. The probability measure δ can be thought of as the sampling distribution used by pure random search. The first sample point X_0 is selected according to δ on S. At each subsequent iteration, the algorithm either hesitates at the current objective function value y with hesitating probability $1 - b(y)$, or strictly improves with probability $b(y)$, where b is a given function from $\mathbb{R} \rightarrow [0, 1]$. If the algorithm strictly improves, it then generates a point according to the normalized restriction of δ on the current improving set. The bettering function b is assumed to be measurable and bounded away from zero on the region of interest. Notice that PAS is the special case of HAS occurring when δ is the uniform distribution on S, and there is constant improvement, i.e., $b \equiv 1$. If the domain S is finite, then HAS with $b \equiv 1$ is the same as *strong* pure adaptive search.

The hesitant adaptive search algorithm is stated more formally below.

Hesitant Adaptive Search (HAS)

Step 0. Initialize $X_0 \in S$ according to probability measure δ on S. Set $k = 0$. Set $Y_0 = f(X_0)$.

Step 1. Generate X_{k+1} according to the normalized restriction of δ on the improving set $S_k = \{x : x \in S \text{ and } f(x) < Y_k\}$ with probability $b(Y_k)$, and otherwise set $X_{k+1} = X_k$. Set $Y_{k+1} = f(X_{k+1})$.

Step 2. If a stopping criterion is met, stop. Otherwise, increment k and return to Step 1.

As in chapter 2, let $N(y)$ be the number of iterations required to first achieve an objective function value of y or less, $y \in \mathbb{R}$. This is also referred to as the number of iterations to convergence, or the number of iterations to land in the termination region, where the termination region T is $(-\infty, y]$. As an aside, open termination regions of the form $(-\infty, y)$ can also be treated using the techniques presented here.

In the development of the distribution of $N(y)$, it is easier to work with $M(y)$ the number of iterations *just before* landing in the termination region. The two measures of performance are easily related; $N(y) = 1 + M(y)$.

The probability measure δ on S dictates the sampling distribution in the domain, and it is convenient to use ρ as the range probability measure defined by $\rho(A) = \delta(f^{-1}[A])$ for each measurable subset A of $\mathbb{R}$. We assume throughout that $\rho(T) > 0$ for $T = (-\infty, y]$; this helps ensure that the expected value and the variance of the number of iterations to convergence are finite. We let p be the range cumulative distribution function associated with ρ, so for $y \in \mathbb{R}$,

$$p(y) = \rho((-\infty, y]) = \delta(S(y)), \text{ where } S(y) = \{x \in S : f(x) \leq y\}.$$

This is consistent with the meaning of $p(y)$ in Chapter 2.

Notice that we are assuming that the course of a run of HAS is completely determined by ρ and b, which depend only on objective function values and are independent of specific points in the domain. Many algorithms do not have this property. For instance, in simulated annealing the local neighborhood of one point in the domain may include the global minimum, while the neighborhood of another point in the domain with the same objective function value might be far from the global minimum. Hence the probability of sampling the global minimum on the next iteration would differ for those two points in the domain. There are also instances where an algorithm on a class of problems depends only on the range and not on the domain. The Improving Hit-and-Run algorithm (discussed in Chapter 6) is analyzed on a class of problems where $p(y)$ and $b(y)$ are independent of the specific point in the domain and are determined by the objective function value.

We proceed now to derive the distribution of the number of HAS iterations to convergence.

2. Number of HAS Iterations to Convergence

The development of the distribution of the number of iterations to convergence is split into two cases: the case of a continuous range distribution termed "continuous HAS," and the case of a discrete range distribution termed "discrete HAS." We then combine these results to produce the main result (Theorem 3.6) describing the distribution of the number of iterations to termination in the general mixed continuous-discrete case, and show (Corollary 3.7) that the expected number of iterations on either a continuous or discrete global optimization problem is

$$E[N(y)] = 1 + \int_{(y,\infty)} \frac{d\rho(t)}{b(t)p(t)}.$$

2.1 Continuous HAS Distribution

The HAS analysis distinguishes between those iterations that make improvement in the objective function, and those iterations that hesitate at a distinct objective function value. Informally, we can compare a run of HAS to the act of leaping down a ladder or sliding down a pole. The heights of the ladder rungs, or position on the pole, correspond to objective function values. If a problem has a finite number of objective function values, then there are a finite number of ladder rungs; a continuous problem corresponds to the pole where it is possible to stop at any height on the pole. The time spent pausing on a ladder rung or clinging to the pole represents hesitation.

We let $Y_{R(k)}$ denote the kth HAS record, i.e., the kth distinct objective function value. This notation is consistent with that used in Chapter 2.

Suppose that it takes T_y distinct objective function values in a run of HAS before the termination level y is reached. These values, termed the "records" of the run, are denoted $Y_{R(1)}, \ldots, Y_{R(T_y)}$. Using the ladder image, $Y_{R(1)}, \ldots, Y_{R(T_y)}$ represent the levels of the rungs used in order to reach level y. On each rung used the algorithm may hesitate, or "mark-time." The number of times it steps on the rung at level $Y_{R(k)}$ is determined by the geometric distribution with parameter $b(Y_{R(k)})$. In order to establish the main result we must first show that for HAS the rung levels $Y_{R(k)}$ are identical to the PAS objective function values, W_k.

The following lemma is similar to Lemma 2.1, and first appeared in [181, Lemma 1]. It shows that the objective function values of PAS are equal in distribution to the record values of HAS.

LEMMA 3.1 *The stochastic process* $\left\{Y_{R(k)},\ k = 0, 1, \ldots\right\}$ *of HAS records is equal in distribution to the stochastic process* $\{W_k,\ k = 0, 1, \ldots\}$ *of PAS, i.e.,*

$$\{W_k\} \sim \left\{Y_{R(k)}\right\}$$

when using the same range probability measure ρ.

Proof: The proof, as in [181], follows the pattern of Lemma 2.1. For x, $y \in \mathbb{R}$ with $y \leq x$, evidently

$$P[Y_{R(k+1)} \leq y | Y_{R(k)} = x] = \frac{p(y)}{p(x)} = P[W_{k+1} \leq y | W_k = x].$$

It is then straightforward to show that the marginal distributions of $Y_{R(k)}$ and W_k coincide, using induction and the equality of the conditional distributions. Consequently the joint distributions are equal. ■

It is technically easier to develop the number of iterations to convergence using the number of iterations *just prior* to achieving the termination region $(-\infty, y]$, $N(y) = 1 + M(y)$. Let M_i be the number of times that the algorithm samples the ith distinct level $Y_{R(i)}$, for $i = 1, \ldots, T_y$. The total number of HAS iterations prior to reaching level y can then be expressed as

$$M(y) = \sum_{i=1}^{T_y} M_i$$

where M_i follows a geometric distribution with the bettering probability $b(Y_{R(i)})$ as the parameter. The main result in this subsection, Theorem 3.2, establishes the factorial moment-generating function of the random variable $M(y)$, so completely determining its distribution. Corollary 3.3 provides the mean and variance for $N(y)$ for continuous HAS.

As shown in Theorem 2.4, PAS iterations in the continuous case form a Poisson process, and intuitively it makes sense that generalizing to include hesitation makes HAS in the continuous case a marked Poisson process: each occurrence of $Y_{R(i)}$ has associated with it a "mark." In this situation the mark is a geometric random variable with parameter $b(Y_{R(i)})$. Such processes have been studied, for example, in [88].

We now provide the moment generating function for $M(y)$, as originally appeared in [181, Theorem 1], allowing a description of the complete distribution of $M(y)$.

THEOREM 3.2 *For continuous HAS, the factorial moment-generating function of $M(y)$ is*

$$E\left[z^{M(y)}\right] = \exp\left\{\int_y^\infty \frac{z-1}{(zb(t)-(z-1))p(t)} d\rho(t)\right\}$$

Proof: The proof is from [181]. The factorial moment-generating function of $M(y) = \sum_{i=1}^{T_y} M_i$ is

$$E\left[z^{M(y)}\right] = E\left[z^{\sum_{i=1}^{T_y} M_i}\right].$$

In order to find an explicit form for this expression, in terms of the range distribution ρ and hesitation parameters $b(Y_{R(i)})$ of HAS, we condition in two stages: first upon T_y, the number of records in (y, ∞) before terminating, then upon $\mathbf{Y} = (Y_{R(1)}, \ldots, Y_{R(T_y)})$, the objective function values of the records. Thus

$$\begin{aligned} E\left[z^{M(y)}\right] &= E_{T_y}\left[E_{\mathbf{Y}}\left[E_{M_i}\left[z^{\sum_{i=1}^{T_y} M_i}|T_y, \mathbf{Y}\right]|T_y\right]\right] \\ &= E_{T_y}\left[E_{\mathbf{Y}}\left[E_{M_i}\left[\prod_{i=1}^{T_y} z^{M_i}|T_y, \mathbf{Y}\right]|T_y\right]\right] \\ &= E_{T_y}\left[E_{\mathbf{Y}}\left[\prod_{i=1}^{T_y} E_{M_i}\left[z^{M_i}|T_y, \mathbf{Y}\right]|T_y\right]\right] \end{aligned}$$

because, given the number of rungs used on the ladder T_y and their locations $\mathbf{Y}$, the M_i are independent. We then have

$$E\left[z^{M(y)}\right] = E_{T_y}\left[E_{\mathbf{Y}}\left[\prod_{i=1}^{T_y} \frac{zb(Y_{R(i)})}{1 - z(1 - b(Y_{R(i)}))}|T_y\right]\right]$$

because $E\left[z^{M_i}\right] = \frac{zb(Y_{R(i)})}{1-z(1-b(Y_{R(i)}))}$ when M_i has a geometric distribution with parameter $b\left(Y_{R(i)}\right)$. Now if T_y is fixed, the locations $Y_{R(1)}, \ldots,$ $Y_{R(T_y)}$ are independent and identically distributed variables in (y, ∞), with common cumulative distribution $1 - m(t)/m(y)$ ([88, page 22]), where $m(t) = \int_t^\infty d\rho(s)/p(s)$ is the mean value function for continuous PAS [29]. Let L denote this common random variable for the locations. Then

$$\begin{aligned} E\left[z^{M(y)}\right] &= E_{T_y}\left[\prod_{i=1}^{T_y} E_L\left[\frac{zb(l)}{1 - z(1 - b(l))}\right]\right] \\ &= E_{T_y}\left[\left(E_L\left[\frac{zb(l)}{1 - z(1 - b(l))}\right]\right)^{T_y}\right] \\ &= \exp\left\{m(y)\left(E_L\left[\frac{zb(l)}{1 - z(1 - b(l))}\right] - 1\right)\right\} \end{aligned}$$

because $E[a^X] = e^{\mu(a-1)}$ when X has a Poisson distribution with parameter μ and $a \in \mathbb{R}$. We have used the fact that T_y has a Poisson distribution with parameter $m(y)$.

Since the cumulative distribution function for location values in (y, ∞) is $F(t) = 1 - m(t)/m(y)$, it follows that $dF(t) = \frac{d\rho(t)}{m(y)p(t)}$. Thus we have

$$E\left[z^{M(y)}\right] = \exp\left\{m(y)\left(\int_y^\infty \frac{zb(t)}{(1 - z(1 - b(t)))} \frac{d\rho(t)}{m(y)p(t)} - 1\right)\right\}$$

$$= \exp\left\{\int_y^\infty \frac{zb(t)}{(1-z(1-b(t)))p(t)}d\rho(t) - m(y)\right\}$$
$$= \exp\left\{\int_y^\infty \frac{z-1}{(zb(t)-(z-1))p(t)}d\rho(t)\right\}$$

the expression for $E\left[z^{M(y)}\right]$ stated in the theorem. ■

The factorial moment-generating function provided by Theorem 3.2 leads to explicit expressions for the factorial moments of all orders. These in turn provide the central moments of all orders through the use of Stirling numbers of the second kind [65]. In this manner we produce a complete description of the distribution of $M(y)$.

The distribution of $M(y)$ is now used to provide the mean and variance of $N(y)$ for continuous HAS in the following corollary. The expression for the mean first appeared in [29, Theorem 1], and the variance appeared in [181, Corollary 1].

COROLLARY 3.3 *For continuous HAS,*

i) the expected value of $N(y)$ *is*

$$E[N(y)] = 1 + E[M(y)] = 1 + \int_y^\infty \frac{d\rho(t)}{b(t)p(t)},$$

ii) the variance of $N(y)$ *is*

$$Var[N(y)] = Var[M(y)] = \int_y^\infty \left(\frac{2}{b(t)} - 1\right)\frac{d\rho(t)}{b(t)p(t)}.$$

Proof: The mean value function used in i) is found by differentiating the factorial moment-generating function and setting z equal to one, as follows

$$\begin{aligned} E\left[M(y)\right] &= \frac{d}{dz}E\left[z^{M(y)}\right]\Big|_{z=1} \\ &= \exp\left\{\int_y^\infty \frac{z-1}{(zb(t)-(z-1))p(t)}d\rho(t)\right\}\Big|_{z=1} \\ &\quad \frac{d}{dz}\int_y^\infty \left\{\frac{z-1}{(zb(t)-(z-1))p(t)}\right\}d\rho(t)\Big|_{z=1} \\ &= \frac{d}{dz}\int_y^\infty \left\{\frac{z-1}{(zb(t)-(z-1))p(t)}\right\}d\rho(t)\Big|_{z=1}. \end{aligned}$$

Note that the conditions of the Lebesgue Dominated Convergence Theorem are satisfied, so that we can differentiate under the integral. Now,

$$\frac{d}{dz}\left\{\frac{z-1}{(zb(t)-(z-1))p(t)}\right\}\Bigg|_{z=1} = \frac{1}{b(t)p(t)},$$

and so

$$E[M(y)] = \int_y^\infty \frac{d\rho(t)}{b(t)p(t)}.$$

Since $N(y) = 1 + M(y)$, we get the result in i).

In similar fashion, we can find $E\left[M(y)(M(y)-1)\right]$ by differentiating $E[z^{M(y)}]$ twice with respect to z and setting $z = 1$. This gives that

$$E\left[M(y)(M(y)-1)\right] = \int_y^\infty \frac{2(1-b(t))\, d\rho(t)}{b(t)^2 p(t)} + \left(\int_y^\infty \frac{d\rho(t)}{b(t)p(t)}\right)^2$$

which can be manipulated to obtain the expression in ii) using routine calculations. ■

2.2 Discrete HAS Distribution

If the domain of the global optimization problem is discrete, then the range cumulative distribution function p is no longer continuous but becomes a step function. Any cumulative distribution function can have at most countably many points of discontinuity, [12, page 176], so we let $\{y_i\}$ denote these points of discontinuity in p and let I^D index these points. Thus the set of points of discontinuity is $\{y_i : i \in I^D\}$.

We describe the number of iterations just prior to convergence as

$$M(y) = \sum_{i \in I^D} M_i$$

where M_i is the number of iterations the sample path spends at level y_i. That is, M_i is zero if we do not visit y_i and equal to the number of steps at y_i if we do visit that rung. The probability that level y_i is visited does not depend on which points above y_i are visited [29] and so we can conclude that the M_i in the above countable sum are mutually independent random variables. For notational brevity in this discrete case we denote $\rho(\{y_i\})$ by π_i, $p(y_i)$ by p_i, consistent with Chapter 2 Section 4.2, and also $b(y_i)$ by b_i, for each $i \in I^D$.

The following theorem presents the distribution of $M(y)$ for discrete HAS, which provides the distribution of $N(y)$ as well. Explicit expressions for the expected value and variance for discrete HAS appear in the following corollary. These results are based on [29, 181].

THEOREM 3.4 *For discrete HAS,* $M(y) = \sum_{i \in I^D} M_i$*, a sum of mutually independent random variables, with*

$$M_i = \begin{cases} 0 & \textit{with probability} \quad 1 - \frac{\pi_i}{p_i} \\ k & \textit{with probability} \quad \frac{\pi_i}{p_i} b_i (1-b_i)^{k-1} \end{cases}.$$

That is, M_i *is a mixture of a degenerate distribution (with all its weight at zero) and a geometric distribution, with parameter* b_i*, by a Bernoulli distribution with parameter* π_i/p_i*.*

Proof: The value of M_i, the number of iterations which the sample path spends at y_i, is determined by the outcome of a Bernoulli random variable with probability of success of π_i/p_i, that is, of visiting level y_i for at least one iteration [29]. If the sample path does not visit y_i, then M_i is zero, which occurs with probability $1 - \pi_i/p_i$. If the sample path does visit y_i, then the number of stays k at level y_i until we move downward is determined by a geometric random variable with probability of bettering (that is, moving downward) of b_i. This proves the theorem. ■

COROLLARY 3.5 *For discrete HAS,*

i) the expected value of $N(y)$ *is*

$$E[N(y)] = 1 + E[M(y)] = 1 + \sum_{i \in I^D} \frac{\pi_i}{b_i p_i},$$

ii) the variance of $N(y)$ *is*

$$Var[N(y)] = Var[M(y)] = \sum_{i \in I^D} \left(\frac{2}{b_i} - \frac{\pi_i}{b_i p_i} - 1 \right) \frac{\pi_i}{b_i p_i}.$$

Proof: Observe that

$$\begin{aligned} E[M_i] &= P[M_i \geq 1]\ E[\text{geometric random variable with parameter } b_i] \\ &= \frac{\pi_i}{p_i} \frac{1}{b_i}, \end{aligned}$$

whence i) follows. It follows routinely that

$$\begin{aligned} E[M_i^2] &= \frac{\pi_i}{p_i} E[(\text{geometric random variable with parameter } b_i)^2] \\ &= \frac{\pi_i (2 - b_i)}{b_i^2 p_i}, \quad \text{so} \\ \text{Var}(M_i) &= \frac{\pi_i (2 - b_i)}{b_i^2 p_i} - \frac{\pi_i^2}{b_i^2 p_i^2} \\ &= \left(\frac{2}{b_i} - \frac{\pi_i}{b_i p_i} - 1 \right) \frac{\pi_i}{b_i p_i} \end{aligned}$$

whence ii) follows, using the mutual independence of the M_i. ∎

The expectation results of Corollary 3.3 i) and Corollary 3.5 i) are analogous. This allows a comparison of whether a continuous problem is harder or easier than a discretized version of the same problem. Using HAS, we see in Section 3, that discretization of the range measure reduces the expected number of iterations to termination (see Table 3.2). The variance expressions of Corollary 3.3 ii) and Corollary 3.5 ii) are almost analogous, differing by a "finite correction factor" of $-\pi_i/(b_i p_i)$; discretization also reduces the variance, as we shall see illustrated later in the examples of Section 3.

Also compare the expectation results for HAS, Corollaries 3.3 i) and 3.5 i), with the expectation results for PAS, Corollary 2.5 i) and Theorem 2.8 i). It is clear that PAS is a special case of HAS with $b \equiv 1$, and the introduction of hesitating with a general bettering probability b falls neatly into the expectation results.

2.3 General HAS Distribution

The analysis of HAS for continuous and discrete cases can be combined to provide a description of the performance of HAS for the general case of mixed continuous-discrete global optimization. In the general case, the range distribution p can be expressed as a convex combination of a continuous range cumulative distribution function and a discrete range distribution function. It would be convenient if $M(y)$, the number of iterations prior to termination in $(-\infty, y]$, could be expressed straightforwardly by adding the number of iterations in the continuous portion $M^C(y)$ to the number of iterations in the discrete portion $M^D(y)$. A little reflection on this possibility, however, leads to the realization that the way in which the measures associated with the continuous and discrete parts interlock, also determines the distribution of $M(y)$. Fortunately, we are still able to use our results for the continuous and discrete cases in the next theorem, which describes the distribution in the general case.

We partition the region prior to termination (y, ∞) with a collection of non-empty open sets where p is continuous, and the points of discontinuity p_i. As in the purely discrete case, we denote the points of discontinuity of p by $\{y_i : i \in I^D\}$, and note there are at most countably many such points. Also let $\pi_i = \rho(\{y_i\})$, $p_i = p(y_i)$ and $b_i = b(y_i)$ for each $i \in I^D$. Note that p is continuous on at most countably many open sets indexed by I^C, $\{U_i : i \in I^C\}$, such that this collection of non-empty open sets and the points of discontinuity form a partition of (y, ∞). Formally, $\bigcup_{i \in I^C} U_i \cup \bigcup_{i \in I^D} \{y_i\} = (y, \infty)$. If the number of discontinuities is finite, with the largest being y_K, then the number of open sets U_i is just the number of discontinuities plus one, and the "top" one, say U_K, is (y_K, ∞). In general, we let $U_i = (a_i, b_i)$. Let $M(U_i)$ denote the number of iterations HAS spends in U_i and $M_i = M(\{y_i\})$ denote the number of iterations HAS spends at y_i.

The following theorem and subsequent corollary present the main result of this chapter with a characterization of the performance of HAS in the general case. The general expression for the expected value first appeared in [29, Theorem 1], and the distribution of HAS first appeared in [181, Theorem 2].

THEOREM 3.6 *Let p be a general range distribution function and $\{\{y_i\} : i \in I^D\}$ and $\{U_i = (a_i, b_i) : i \in I^C\}$ be the associated partition of (y, ∞). Then for HAS with bettering function b,*

$$M(y) = \sum_{i \in I^C} M(U_i) + \sum_{i \in I^D} M_i \ ,$$

an at most countable sum of mutually independent random variables, where

i) $M(U_i)$ has factorial moment-generating function

$$E\left[z^{M(U_i)}\right] = \exp\left\{\int_{(a_i,b_i)} \frac{z-1}{(zb(t)-(z-1))p(t)} d\rho(t)\right\}, \qquad \text{and}$$

ii) M_i has distribution given by

$$M_i = \begin{cases} 0, & \text{with probability } 1 - \frac{\pi_i}{p_i} \\ k, & \text{with probability } \frac{\pi_i}{p_i} b_i (1-b_i)^{k-1}. \end{cases}$$

Proof: As in [29] and [181], the proof notes that the component random variables are mutually independent. The levels assumed by HAS in U_i form a Poisson process and so a slight modification of the argument in Theorem 3.2 suffices to show i). The probability that general HAS strikes $\{y_i\}$ is π_i / p_i, again appealing to [29]. The distribution of M_i thus follows by including the geometric distribution using the argument of Theorem 3.4. ∎

COROLLARY 3.7 *For general HAS,*

i) the expected value of $N(y)$ is

$$E[N(y)] = 1 + E[M(y)]$$

$$= 1 + \sum_{\{U_i : i \in I^C\}} \int_{U_i} \frac{d\rho(t)}{b(t)p(t)} + \sum_{\{y_i : i \in I^D\}} \frac{\pi_i}{b_i p_i}$$

$$= 1 + \int_{(y,\infty)} \frac{d\rho(t)}{b(t)p(t)},$$

ii) the variance of $N(y)$ *is*

$$\begin{aligned} Var[N(y)] &= Var[M(y)] \\ &= \sum_{\{U_i : i \in I^C\}} \int_{U_i} \left(\frac{2}{b(t)} - 1\right) \frac{d\rho(t)}{b(t)p(t)} \\ &\quad + \sum_{\{y_i : i \in I^D\}} \left(\frac{2}{b_i} - \frac{\pi_i}{b_i p_i} - 1\right) \frac{\pi_i}{b_i p_i}. \end{aligned}$$

Proof: The proof follows that given in [181]. For i), note that

$$E[M(y)] = \sum_{i \in I^C} E[M(U_i)] + \sum_{i \in I^D} E[M_i].$$

Now

$$E[M(U_i)] = \int_{U_i} \frac{d\rho(t)}{b(t)p(t)}$$

using the factorial moment-generating function of $M(U_i)$ given in Theorem 3.6 i) and the argument in the proof of Corollary 3.3 i). Also $E[M_i] = \pi_i/(b_i p_i)$ as in the proof of Corollary 3.5 i). Since the disjoint partition $\{U_i : i \in I^C\} \cup \{\{y_i\} : i \in I^D\}$ covers (y, ∞), the result in i) follows.

For ii), note that $\text{Var}[M(U_i)] = \sum_{i \in I^C} \text{Var}[M(U_i)] + \sum_{i \in I^D} \text{Var}[M_i]$ since the component variables are mutually independent. Now

$$\text{Var}[M(U_i)] = \int_{U_i} \left(\frac{2}{b(t)} - 1\right) \frac{d\rho(t)}{b(t)p(t)}$$

using the factorial moment-generating function for $M(U_i)$ given in Theorem 3.6 i) and the argument in the proof of Corollary 3.3 ii). Also $\text{Var}[M_i] = \left(\frac{2}{b_i} - \frac{\pi_i}{b_i p_i} - 1\right) \frac{\pi_i}{b_i p_i}$ using Theorem 3.6 ii) and the proof of Corollary 3.5 ii), whence ii) follows. ∎

All expectations and variances of $M(y)$ and $N(y)$ calculated throughout are finite. This follows from the assumption that the probability of successfully landing in the terminating region $(-\infty, y]$ is positive, i.e., the probability of success $\rho((-\infty, y]) > 0$, and that the bettering probability is bounded away from zero. Thus $\int_{(y,\infty)} \frac{d\rho(t)}{p(t)}$ is finite whence $\int_{(y,\infty)} \frac{d\rho(t)}{b(t)p(t)}$ is finite, which in turn can be used to show the variances computed are finite.

The main result of this chapter is captured by the distribution for HAS on a general mixed continuous-discrete global optimization problem. The expressions show the role of the underlying sampling distribution and bettering probability on the performance of HAS. If the bettering probability function has a lower bound of $B > 0$, then

$$E[N(y)] = 1 + \frac{1}{B} \int_{(y,\infty)} \frac{d\rho(t)}{p(t)}$$

provides an upper bound on the number of HAS iterations until convergence.

3. Numerical Examples of HAS

In this section we present two numerical examples, and simulate HAS using MATLAB to illustrate its rate of convergence. The first example illustrates the link between a continuous problem and its discrete counterpart; the second example involves a simulation of a mixed continuous-discrete domain problem. In each case we compare the sample mean and variance with the theoretical predictions. The examples are simple, but suffice to illustrate the working of HAS. These examples were developed by Kristinsdottir, et al. [93, 186].

Example 1. We illustrate the distribution of $N(y)$ for HAS applied to a linear objective function on a continuous interval domain and then a discrete domain. The two forms of the problem are:

$$\begin{array}{lll} \min f(x) = x & \text{for } x \in [0, 10] & \text{(continuous)} \\ \min f(x) = x & \text{for } x \in \{1, 2, \ldots, 10\} & \text{(discrete)} \end{array}$$

In both cases a uniform distribution δ on the domain is used, so projecting this onto the range of f in each case we have

$$\begin{array}{lll} p(y) = y/10 \text{ and } d\rho(y) = dy/10 & \text{for } y \in [0, 10] & \text{(continuous)} \\ p_i = i/10 \text{ and } \pi_i = 1/10 & \text{for } i \in \{1, 2, \ldots, 10\} & \text{(discrete)} \end{array}$$

A linear hesitation function b is used, so that the probability of bettering is smaller for objective function values closer to the optimum:

$$\begin{array}{lll} b(y) = y/10 & \text{for } y \in [0, 10] & \text{(continuous)} \\ b_i = i/10 & \text{for } i \in \{1, 2, \ldots, 10\} & \text{(discrete)} \end{array}$$

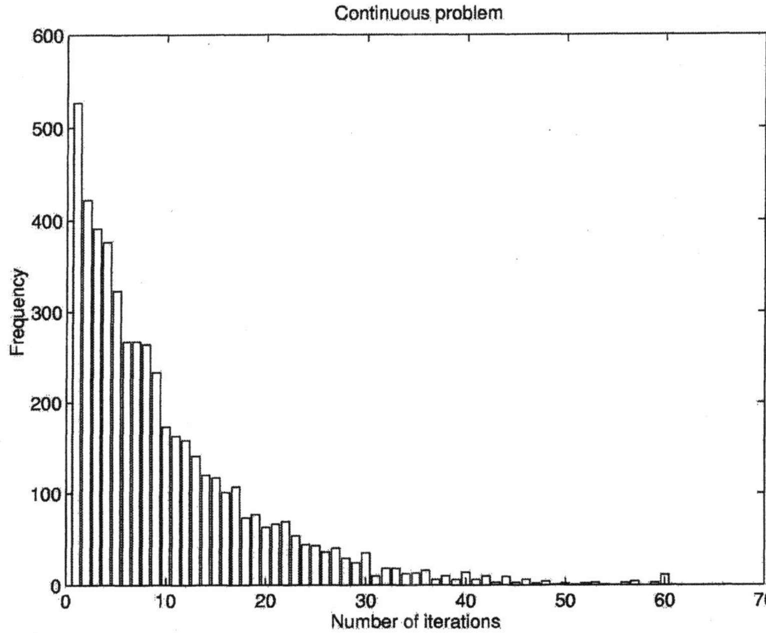

Figure 3.1. Distribution of the number of iterations to reach $(-\infty, 1]$ for the continuous optimization problem of Example 1.

In both cases, HAS was run 5000 times and each run terminated when an iteration first reached $(-\infty, 1]$ in the continuous case, and $\{1\}$ in the discrete case. Figures 3.1 and 3.2 show the frequency histograms for the number of iterations to convergence, $N(1) = 1 + M(1)$, for the continuous and discrete cases respectively. Notice that the histogram for the continuous case resembles a Poisson distribution and that the histogram for the discrete case has lower mean and variance than in the continuous case.

The mean and variance for the number of iterations to convergence for the continuous and discrete cases, using Corollaries 3.3 and 3.5, are

i) continuous case

$$E[N(1)] = 1 + \int_1^{10} \frac{d\rho(t)}{b(t)p(t)} = 10$$

$$\text{Var}[N(1)] = \int_1^{10} \left(\frac{2}{b(t)} - 1\right) \frac{d\rho(t)}{b(t)p(t)} = 90$$

ii) discrete case

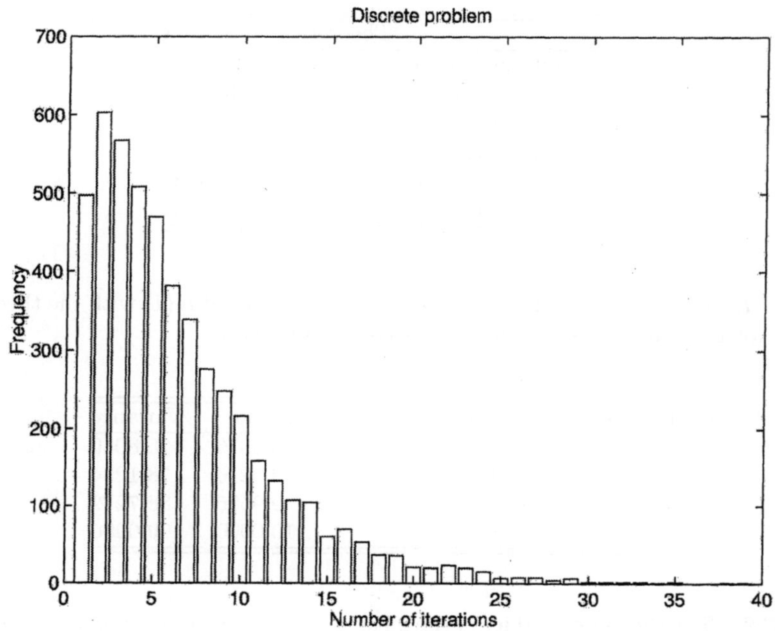

Figure 3.2. Distribution of the number of iterations to reach $\{1\}$ for the discrete optimization problem of Example 1.

$$E[N(1)] = 1 + \sum_{i=2}^{10} \frac{\pi_i}{b_i p_i} = 6.5$$

$$\text{Var}[N(1)] = \sum_{i=2}^{10} \left(\frac{2}{b_i} - \frac{\pi_i}{b_i p_i} - 1 \right) \frac{\pi_i}{b_i p_i} = 25.8$$

The sample mean for the simulation with 5000 runs on the continuous case of Example 1 is 10.0772 and the sample variance is 90.7406, showing close agreement with the theoretical values as summarized in Table 3.1. Similarly, for the discrete case the sample mean is 6.4384 and the sample variance is 26.3391, again closely matching the theoretical values, illustrated in Table 3.1.

The analysis of HAS provides a mechanism to compare performance on similar continuous and discrete problems. Table 3.2 shows how the values for the discrete case move to those for the continuous case as the refinement increases. The extreme rows in the table correspond to the two cases just discussed. As indicated in the table, as the level of discretization decreases, the performance of HAS on a discrete problem approaches performance of HAS on the limiting continuous problem.

	Theoretical	Sample
i) continuous case		
mean	10	10.0772
variance	90	90.7406
ii) discrete case		
mean	6.5	6.4384
variance	25.8	26.3391

Table 3.1. Comparison of the sample mean and variance of $N(1)$ with the theoretical values for the continuous and discrete cases in Example 1.

Distance between domain points	Mean	Variance
1	6.4976	25.8050
0.1	9.5216	78.1234
0.01	9.9506	88.7272
Continuous	10.0000	90.0000

Table 3.2. The theoretical values of the mean and variance of the number of iterations to convergence $N(1)$ in Example 1, as the space between domain points in $[0, 10]$ decreases.

Example 2. We now examine the performance of HAS for the problem

$$\begin{array}{lrcl} \min & f(x) & = & x \\ \text{for} & x & \in & (0,1) \cup [2,3) \cup [4,5) \cup [6,7) \cup [8,9) \cup \{10\} \text{ (mixed)} \end{array}$$

where the sampling distribution δ places atoms of probability of 0.1 at each of the points 2, 4, 6, 8 and 10, and a probability of 0.1 is placed uniformly in each of the open intervals $(0, 1)$, $(2, 3)$, $(4, 5)$, $(6, 7)$ and $(8, 9)$. The induced range cumulative distribution function, $p(y)$, is pictured in Figure 3.3. We again set $b(y) = y/10$ for $y \in [0, 10]$ and use $(-\infty, 1]$ as the termination region.

HAS was again run 5000 times; Figure 3.4 shows the frequency histogram for $N(1) = 1 + M(1)$, the number of iterations required to reach termination in $(-\infty, 1]$. The equations given in Corollary 3.7 are used to calculate the theoretical mean and variance for $N(1)$ as 7.2 and 33.9 respectively. The sample mean is 7.3862 and the sample variance 36.1563, again closely matching the theoretical values.

4. Combination of PRS and PAS, *(1-p)*PRS+*p*PAS

The HAS algorithm can be used to model various algorithms and gain useful insights. In Chapter 2, the PRS and PAS algorithms were

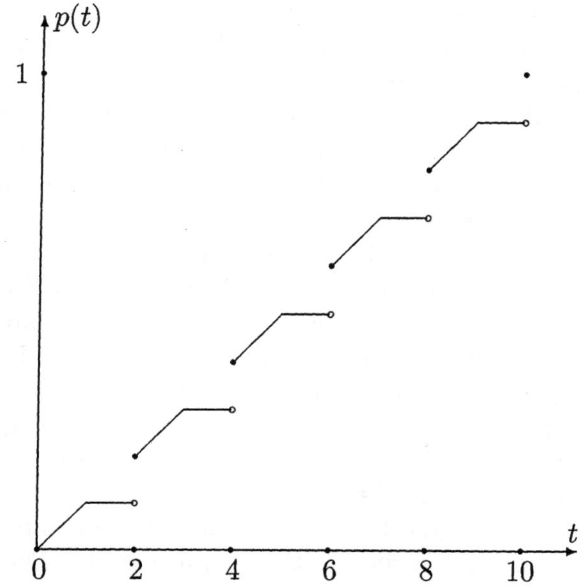

Figure 3.3. The range cumulative distribution function p for the mixed continuous-discrete problem of Example 2. The termination region is $(-\infty, 1]$.

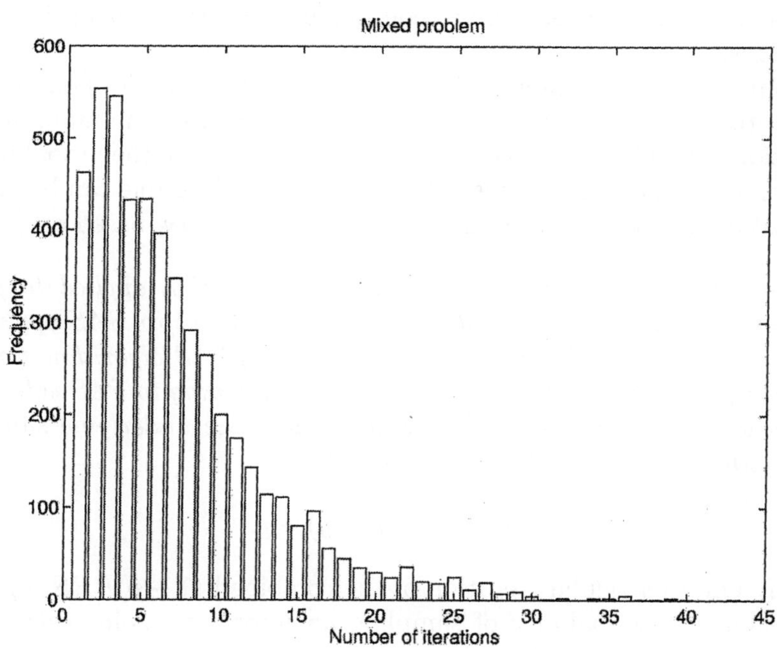

Figure 3.4. Distribution of the number of iterations to reach $(-\infty, 1]$ for the mixed continuous discrete optimization problem of Example 2.

introduced and analyzed as two extreme versions of a random search algorithm. The performance of PAS was shown to be linear in dimension, as compared with the exponential performance of PRS, which suggests the value of making steady improvement. However, in practicality algorithms cannot ensure consistent improvement. This leads us to ask, what is the performance of an algorithm that only produces improving points a fraction of the time?

In an attempt to gain insight into the power of steady improvement, we analyze an algorithm that is a combination of PRS and PAS. The combined PRS-PAS algorithm has a probability p of sampling according to PAS and probability $1-p$ of sampling according to PRS where $0 \leq p \leq 1$. When $p = 1$ the combined PRS-PAS algorithm reduces to PAS with steady improvement, and when $p = 0$ it is simply PRS. An analysis of the combined $(1-p)$PRS+pPAS algorithm in terms of p provides a way to quantify the value of steady improvement.

We use HAS to analyze the performance of the combined PRS-PAS algorithm by choosing a specific bettering probability. The original analysis of the combined PRS-PAS algorithm was performed for a discrete problem using Markov chain analysis by Zabinsky and Kristinsdottir [186]. Using HAS allows a more general analysis of the combined PRS-PAS algorithm for general mixed continuous-discrete global optimization problems.

Even though the combined PRS-PAS algorithm samples points in the non-improving region as well as the improving region, the sampling distribution depends only on the record values, and hence the probability of landing in the non-improving region due to a PRS sample can be viewed as hesitating. This is stated in the following corollary, based on [186].

COROLLARY 3.8 *The combined PRS-PAS algorithm applied to global optimization problem (P) is stochastically equivalent to HAS applied to the same problem (P), where the $(1-p)PRS+pPAS$ algorithm samples according to PRS with probability $1-p$ and according to PAS with probability p, for $0 \leq p \leq 1$, and HAS uses the same underlying sampling distribution $p(y)$ with bettering probability*

$$b(y) = (1-p)p(y) + p.$$

Proof: On a particular iteration of the combined $(1-p)PRS + pPAS$ algorithm, the probability of sampling an improving point using PRS is $p(y)$, and the probability of sampling an improving point using PAS is 1. Since PRS is used with probability $(1-p)$ and PAS is used with probability p, the probability that the combined algorithm makes an improvement is $(1-p)p(y) + p$. Consequently, HAS with a bettering probability of

$$b(y) = (1-p)p(y) + p$$

is equivalent to the combined PRS-PAS algorithm when using the same underlying sampling distribution. ■

The equivalence of the combined PRS-PAS algorithm with HAS using $b(y) = (1-p)p(y) + p$ provides the distribution of $N(y)$ as well as the mean and variance on the number of iterations until convergence. We next explore the impact of p on the performance of the combined PRS-PAS algorithm for the separate continuous case and discrete case.

4.1 Continuous PRS and PAS Combination

The expected number of iterations to termination for the combined PRS-PAS algorithm in the continuous case is stated in the following corollary, identifying the impact of steady improvement on performance.

COROLLARY 3.9 *The expected number of* $(1-p)PRS{+}pPAS$ *iterations until termination is*

$$E[N(y)] \quad = \quad 1 + \frac{1}{p}\ln\left(1 - p + \frac{p}{p(y)}\right)$$

for $0 < p \leq 1$.

Proof: Combining the expected number of iterations for continuous HAS as given in Corollary 3.3 with a specific bettering probability of $b(y) = (1-p)p(y) + p$ from Corollary 3.8, we have

$$\begin{aligned} E[N(y)] &= 1 + \int_y^\infty \frac{d\rho(t)}{b(t)p(t)} \\ &= 1 + \int_y^\infty \frac{d\rho(t)}{((1-p)p(t)+p)p(t)}. \end{aligned}$$

Making a change of variables, $u = p(t)$, and hence changing the limits to $p(\infty) = 1$ and $p(y)$ gives

$$\begin{aligned} E[N(y)] &= 1 + \int_{p(y)}^1 \frac{du}{(1-p)u^2 + pu} \\ &= 1 + \frac{1}{p}\ln\left(\frac{(1-p)u}{(1-p)u+p}\right)\Bigg|_{p(y)}^1 \\ &= 1 + \frac{1}{p}\ln\left(\frac{(1-p)p(y)+p}{p(y)}\right) \end{aligned}$$

$$= \quad 1 + \frac{1}{p} \ln \left(1 - p + \frac{p}{p(y)} \right).$$

■

When $p = 1$, the algorithm reduces to the PAS algorithm. Letting $p = 1$ gives the expected number of iterations as, $E[N(y)] = 1 + ln[1/p(y)]$ which reproduces the result in Corollary 2.5. For the special case when $p = 0$, we have the PRS algorithm. Taking the limit as p approaches zero yields

$$\begin{aligned}
\lim_{p \to 0} E[N(y)] &= \lim_{p \to 0} \quad 1 + \frac{1}{p} \ln \left(1 - p + \frac{p}{p(y)} \right) \\
&= \lim_{p \to 0} \quad 1 + \frac{\frac{d}{dp} \left(ln(1 - p + \frac{p}{p(y)}) \right)}{\frac{d}{dp}(p)} \\
&= \lim_{p \to 0} \quad 1 + \frac{\left(\frac{1}{1-p+\frac{p}{p(y)}} \right) \left(-1 + \frac{1}{p(y)} \right)}{1} \\
&= \frac{1}{p(y)}
\end{aligned}$$

which reproduces the result in Equation 2.2. The following corollary shows that the probability of steady improvement p for the combined PRS-PAS can be simply entered into the bound on expected number of iterations until convergence.

COROLLARY 3.10 *The expected number of iterations get below a value of y for the combined PRS-PAS algorithm, $(1 - p)PRS$+$pPAS$, on a global optimization problem (P) over a convex feasible region S in n dimensions with diameter D and Lipschitz constant at most K, is bounded above as follows*

$$E[N(y)] \quad \leq \quad 1 + \frac{n}{p} \ln \left(\frac{KD}{y - y_*} \right)$$

for $y_* \leq y \leq y^*$ and $0 < p \leq 1$.

Proof: For the combined PRS-PAS algorithm on problem (P), we have from Theorem 2.9 that $p(y) \geq (y - y_*/KD)^n$ and hence, $1/p(y) \leq (KD/y - y_*)^n$, and from Corollary 3.9, we have

$$E[N(y)] \quad = \quad 1 + \frac{1}{p} \ln \left(1 - p + \frac{p}{p(y)} \right)$$

$$\leq \quad 1+\frac{1}{p}\ln\left(1-p+p\left(\frac{KD}{y-y_*}\right)^n\right).$$

Using $1-p+p\left(\frac{KD}{y-y_*}\right)^n \leq (1-p)\left(\frac{KD}{y-y_*}\right)^n + p\left(\frac{KD}{y-y_*}\right)^n = \left(\frac{KD}{y-y_*}\right)^n$ gives

$$\begin{aligned} E[N(y)] &\leq 1+\frac{1}{p}\ln\left(\frac{KD}{y-y_*}\right)^n \\ &\leq 1+\frac{n}{p}\ln\left(\frac{KD}{y-y_*}\right). \end{aligned}$$

■

From the above result, we conclude that the expected number of combined PRS-PAS iterations grows linearly in dimension as long as the probability of steady improvement is positive, $p > 0$. Figure 3.5 plots the expected number of iterations to convergence using a fixed $KD/(y-y_*)$ arbitrarily chosen to be a constant equal to 1.8. It is interesting to see that the expected number of iterations changes slowly as p changes. This shows that one only needs a small probability of sampling in the improving region in order to dramatically improve performance. Also, there is a diminishing return as p exceeds 0.5 and gets closer to 1.0.

4.2 Discrete PRS and PAS Combination

We again explore the performance of the combined PRS-PAS algorithm for the discrete case, and now PAS has two versions, a strong and weak version, as described in Chapter 2. A similar relationship between the combined PRS-PAS algorithm and HAS holds for the discrete case, but now the bettering probability changes slightly when describing weak PAS. The following corollary states the expected number of iterations for the combined PRS-PAS algorithm in both versions on the discrete case.

COROLLARY 3.11 *The expected number of* $(1-p)PRS+pPAS$ *iterations until termination for discrete problems is given as follows,*

$$\begin{aligned} &i) \quad E[N(y)] = 1+\sum_{i\in I^D}\frac{\pi_i}{p_i((1-p)p_{i-1}+p)} \quad \textit{for strong PAS} \\ &ii) \quad E[N(y)] = 1+\sum_{i\in I^D}\frac{\pi_i}{p_{i-1}((1-p)p_i+p)} \quad \textit{for weak PAS} \end{aligned}$$

for $0 \leq p \leq 1$.

Proof: The bettering probability b_i, is the probability of HAS making a strict improvement. On a PRS iteration, the probability of generating a

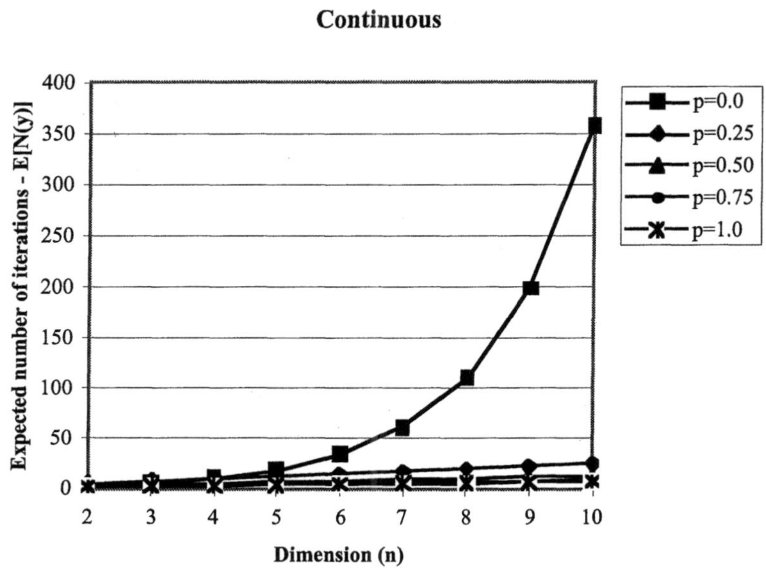

Figure 3.5. The expected number of iterations for the $(1-p)$PRS+pPAS algorithm with several values of p between 0 and 1, and $n = 2, \ldots, 10$.

strictly improving point is p_{i-1}, and this occurs with a probability $1-p$. On an iteration of strong PAS, the probability of generating a strictly improving point is 1, but it is $\frac{p_{i-1}}{p_i}$ for weak PAS, and this occurs with a probability p. Hence, we get the following bettering probabilities,

$$\begin{aligned} i) \quad b_i &= (1-p)p_{i-1} + p && \text{for strong PAS} \\ ii) \quad b_i &= (1-p)p_{i-1} + p\left(\frac{p_{i-1}}{p_i}\right) && \text{for weak PAS.} \end{aligned}$$

Using the expected number of iterations until termination for HAS in the discrete case given in Corollary 3.5, $E[N(y)] = 1 + \sum_{i \in I^D} \frac{\pi_i}{b_i p_i}$, gives

$$\begin{aligned} i) \quad E[N(y)] &= 1 + \sum_{i \in I^D} \frac{\pi_i}{p_i((1-p)p_{i-1} + p)} && \text{for strong PAS} \\ ii) \quad E[N(y)] &= 1 + \sum_{i \in I^D} \frac{\pi_i}{p_{i-1}((1-p)p_i + p)} && \text{for weak PAS.} \end{aligned}$$

■

In order to get analogous results to the linearity results for continuous problems for discrete problems, we consider a domain S that consists of the vertices of an n-dimensional lattice $\{1,\ldots,K\}^n$ with distinct objective function values at all points. The number of distinct objective function values equals the number of points in the lattice, $M = K^n$. Assuming that the underlying distribution is uniform on the domain, and because each point in the domain has unique objective function value, the distribution is also uniform on the range, $\pi_j = 1/M = 1/K^n$ for all $j = 1,\ldots,M$.

COROLLARY 3.12 *The expected number of iterations of the combined* $(1-p)$*PRS+pPAS algorithm with a uniform distribution on a finite global optimization problem with M distinct objective function values is*

i) $E[N(y)] = 1 + M\sum_{j=2}^{M} \frac{1}{ipM+i(1-p)(i-1)}$ *for strong PAS*

ii) $E[N(y)] = \frac{M}{M-1} + M\sum_{j=1}^{M-2} \frac{1}{(M-j-1)(M-j(1-p))}$ *for weak PAS.*

Proof: Assuming a uniform distribution, $p_i = \frac{i}{M}$ and $\pi_i = \frac{1}{M}$, and applying Corollary 3.11 gives, for strong PAS,

$$E[N(y)] = 1 + M\sum_{i=2}^{M} \frac{1}{(ipM + i(1-p)(i-1))}.$$

For weak PAS, we have

$$\begin{aligned} E[N(y)] &= 1 + M\sum_{i=2}^{M} \frac{1}{(i-1)(i(1-p)+Mp)} \\ &= 1 + \frac{1}{M-1} + M\sum_{i=2}^{M-1} \frac{1}{(i-1)(i(1-p)+Mp)} \end{aligned}$$

and using $i = M - j$ thus $j = M - i$ and $j = 1,..,M-2$ gives

$$= \frac{M}{M-1} + M\sum_{j=1}^{M-2} \frac{1}{(M-j-1)(M-j(1-p))}.$$

■

COROLLARY 3.13 *Upper bounds on the expected number of iterations of* $(1-p)$*PRS+pPAS with a uniform distribution on a finite global optimization problem with M distinct objective function values are as follows*

i) $E[N(y)] \leq 1 + \frac{1}{p}\ln M$ *for strong PAS*

ii) $E[N(y)] \leq 1 + \frac{M}{pM+(1-p)}(1+\ln(M-1))$ *for weak PAS* *and assuming an n dimensional lattice, $\{1,\ldots,K\}^n$, with distinct objective function values at all points, the expected number of iterations is bounded as follows,*

$$\begin{array}{llll} i) & E[N(y)] & \leq & 1+\frac{n}{p}\ln K & \textit{for strong PAS} \\ ii) & E[N(y)] & \leq & 1+(1/p)+\frac{n}{p}\ln K & \textit{for weak PAS.} \end{array}$$

Proof: i) Assuming strong PAS

$$\begin{aligned} E[N(y)] &= 1+M\sum_{i=2}^{M}\frac{1}{ipM+i(1-p)(i-1)} \\ &\leq 1+M\sum_{i=2}^{M}\frac{1}{Mip} \\ &= 1+\frac{1}{p}\sum_{i=2}^{M}\frac{1}{i} \\ &= \frac{p-1}{p}+\frac{1}{p}\sum_{i=1}^{M}\frac{1}{i} \\ &\leq \frac{p-1}{p}+\frac{1}{p}(1+\ln M) \\ &= 1+\frac{1}{p}\ln M. \end{aligned}$$

ii) Assuming weak PAS

$$\begin{aligned} E[N(y)] &= \frac{M}{M-1}+M\sum_{j=1}^{M-2}\frac{1}{(m-j-1)(m-j(1-p))} \\ &= \frac{M}{M-1}+M\sum_{j=1}^{M-2}\frac{1}{(m-j-1)(pm+(m-j)(1-p))} \\ &= \frac{M}{M-1}+M\sum_{j=2}^{M-1}\frac{1}{(m-j)(pm+(m-j+1)(1-p))} \\ &= \frac{M}{M-1} \\ &\quad +M\sum_{j=2}^{M-1}\frac{1}{(m-j)pM+(m-j)^2(1-p)+(m-j)(1-p)} \\ &\leq \frac{M}{M-1}+M\sum_{j=2}^{M-1}\frac{1}{(m-j)pM+(m-j)(1-p)} \end{aligned}$$

$$
\begin{aligned}
&= \frac{M}{M-1} + M\sum_{j=1}^{M-2}\frac{1}{j(pM+(1-p))} \\
&= \frac{M}{M-1} + \frac{M}{pM+(1-p)}\left(\sum_{j=1}^{M-2}\frac{1}{j} + \frac{1}{M-1} - \frac{1}{M-1}\right) \\
&= \frac{M}{M-1} - \frac{M}{(M-1)(pM+(1-p))} + \frac{M}{pM+(1-p)}\sum_{j=1}^{M-1}\frac{1}{j} \\
&= \frac{pM}{pM+(1-p)} + \frac{M}{pM+(1-p)}\sum_{j=1}^{M-1}\frac{1}{j} \\
&\leq 1 + \frac{M}{pM+(1-p)}\sum_{j=1}^{M-1}\frac{1}{j} \\
E[N(y)] &\leq 1 + \frac{M}{pM+(1-p)}(1+\ln(M-1)).
\end{aligned}
$$

Assuming an n-dimensional lattice with $M = K^n$ gives, for strong PAS

$$E[N(y)] \leq 1 + \frac{n}{p}\ln K$$

and for weak PAS the bound is

$$
\begin{aligned}
E[N(y)] &\leq 1 + \frac{K^n}{pK^n+(1-p)}(1+\ln(K^n-1)) \\
&\leq 1 + \frac{1}{p}(1+\ln K^n).
\end{aligned}
$$

■

Results for PAS and PRS can be derived as special cases by setting $p = 1$ and $p = 0$ in the corollaries above. Setting $p = 1$ in Corollary 3 gives

$$
\begin{aligned}
&i)\quad E[N(y)] = 1 + \textstyle\sum_{i\in I^D}\frac{\pi_i}{p_i} \quad \text{for strong PAS} \\
&ii)\quad E[N(y)] = 1 + \textstyle\sum_{i\in I^D}\frac{\pi_i}{p_{i-1}} \quad \text{for weak PAS}
\end{aligned}
$$

which agrees with Theorem 2.8.

Assuming a uniform distribution with $p_i = \frac{i}{M}$ and $\pi_i = \frac{1}{M}$ gives,

$$
\begin{aligned}
i)\quad E[N(y)] &= 1 + \textstyle\sum_{j=2}^{M}\frac{1}{j} = \sum_{i=1}^{M}\frac{1}{i} \quad \text{for strong PAS} \\
E[N(y)] &\leq 1 + \ln M
\end{aligned}
$$

$$
\begin{aligned}
ii)\quad E[N(y)] &= 1 + \textstyle\sum_{j=2}^{M}\frac{1}{(j-1)} = 1 + \sum_{i=1}^{M-1}\frac{1}{i} \quad \text{for weak PAS} \\
E[N(y)] &\leq 2 + \ln(M-1)
\end{aligned}
$$

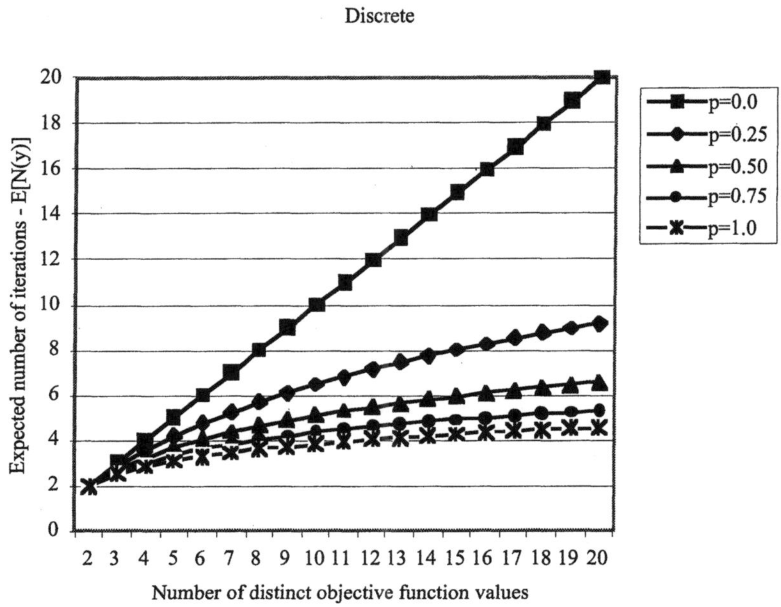

Figure 3.6. Expected number of iterations to convergence for combined PRS-PAS for values of p.

which agrees with the results in Theorem 2.8 and Corollary 2.12.

As in the continuous case, we can see the impact that steady improvement has on the performance of the combined PRS-PAS algorithm by varying p. Figure 3.6 plots the expected number of iterations to convergence using Corollary 3.12 for M ranging from 2 to 20. In this plot, PRS is the line with $p = 0$ and PAS is the line with $p = 1$. The plot illustrates that only a small probability of improvement, such as $p = 0.25$, is enough to dramatically improve performance.

5. Summary

This chapter has presented the distribution, with expressions for the mean and variance, of the number of iterations until convergence for HAS on a continuous, discrete and general range cumulative distribution function. This encompasses global optimization problems with both continuous and discrete variables, and a general underlying sampling distribution. Numerical examples and a combined PRS-PAS algorithm illustrate the power of the analysis of HAS, and demonstrate the impact of the underlying sampling distribution and probability of improvement on performance. HAS is more realistic than PAS because it allows hesitation, and thus is a step closer towards understanding the complexity of more general random search algorithms.

Chapter 4

ANNEALING ADAPTIVE SEARCH

While pure adaptive search and hesitant adaptive search use an underlying sampling distribution that is restricted to sampling from *improving sets,* annealing adaptive search (AAS) always samples from the *original feasible region* but modifies the sampling distribution so that the probability of sampling in the improving level set is high. The record values of AAS were called adaptive search and analyzed by Romeijn and Smith [139]; they were shown to stochastically dominate PAS. Hence the number of improving points of AAS inherits the linear complexity of PAS. The number of sample points (including non-improving points) of AAS can be analyzed using a relationship to HAS. The additional samples needed before achieving an improving point may be interpreted as hesitating, and the link to HAS provides a bound on the expected number of sample points of AAS to achieve a specified level of accuracy. The name *annealing adaptive search* (suggested in [180]) is used because the algorithm samples according to a Boltzmann distribution where the temperature parameter is updated at each improving point. The algorithm is theoretical in the sense that it assumes points have an exact Boltzmann distribution, but the purpose of the analysis is to gain understanding of simulated annealing and other stochastic algorithms.

Annealing adaptive search is defined in the following section, and compared to the definition of adaptive search. Analysis of AAS on continuous and finite global optimization problems is presented in Section 2, including the theoretical linear performance of record values [139]. An upper bound on the number of sample points of AAS is also given [154], using a special case of HAS. Furthermore an adaptive cooling schedule is discussed in Section 3. The cooling schedule was originally derived by Romeijn and Smith [139] and extended by Kiatsupaibul and Smith

[85, 86]. This analysis provides a theoretical basis for a cooling schedule that can be used when implementing a simulated annealing algorithm. The challenge remains to efficiently generate sample points according to a Boltzmann distribution, and an approach to approximate this using Hit-and-Run is discussed in Chapter 6.

1. Annealing Adaptive Search (AAS)

Annealing adaptive search is defined on a global optimization problem (P), and it samples points according to a family of Boltzmann distributions, which become more concentrated around the global optimum as the temperature decreases. The Boltzmann family is parameterized by a temperature parameter, and the distribution is denoted π_T. In our context with objective function $f(x)$ and feasible region S, the density for the Boltzmann distribution with parameter T is,

$$g_T(x) = \frac{e^{-f(x)/T}}{\int_S e^{-f(z)/T} dz} \tag{4.1}$$

where f is the objective function of the underlying optimization problem, S is the feasible region, and $T > 0$. Annealing adaptive search (based on adaptive search introduced in [139]) updates the temperature parameter every time an improving point (record value) is found.

To develop an intuitive understanding of AAS, it is important to review the Boltzmann distribution. When T is ∞, the Boltzmann distribution is equivalent to the uniform distribution, so the initial step of AAS is to simply generate a point uniformly distributed on S. The Boltzmann distribution reflects the shape of the objective function, and as T approaches zero, the density peaks at the global optimum, hence points generated from a Boltzmann distribution with a low temperature value are more likely to be close to the global optimum.

Consider a simple quadratic example, where S is the interval [0,3] and $f(x)$ is $(x-1)^2$, so the optimum is at $x_* = 1$ with $y_* = 0$. Figure 4.1 illustrates a family of Boltzmann density functions and cumulative distributions with three temperature values, $T = 0.1, 1.0$, and 10.0. When T is 10, the density looks nearly flat, which is close to a uniform distribution. At $T = 0.1$, the density favors points close to the optimum.

The Boltzmann cumulative distribution for the same quadratic function and values of T are illustrated in the bottom graph in Figure 4.1. The curve for $T = 0.1$ is always greater than the curve for $T = 1$, which is also greater than the curve for $T = 10$. This is a property of the Boltzmann family of distributions, and can be formalized with the concept of stochastic dominance. We use the following definition of stochastic dominance, as in Ross [142].

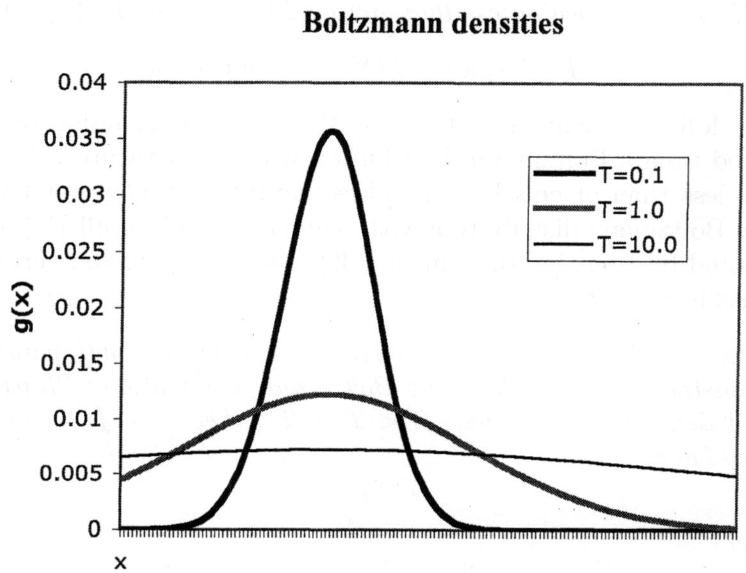

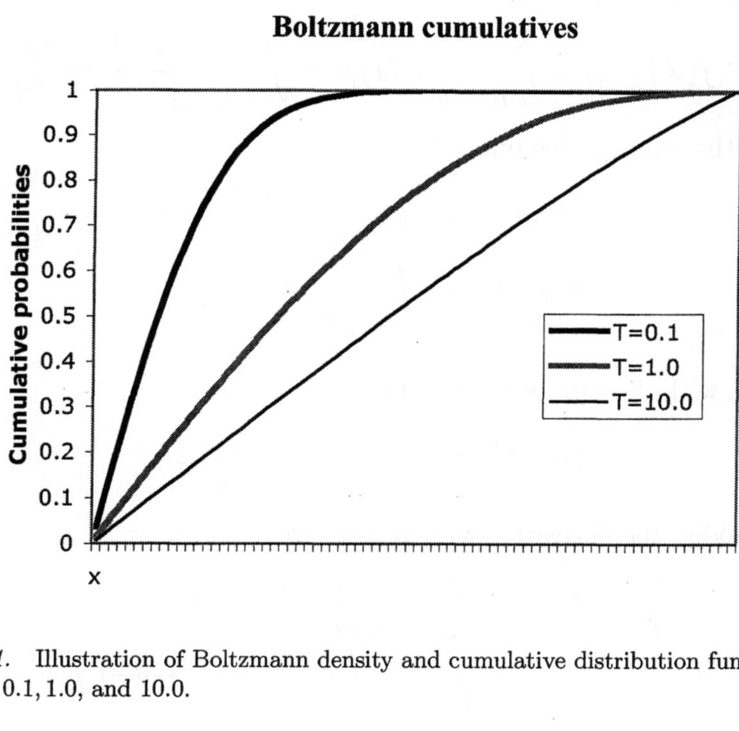

Figure 4.1. Illustration of Boltzmann density and cumulative distribution functions with $T = 0.1, 1.0$, and 10.0.

DEFINITION 4.1 *Let X and Y be real-valued random variables, we say that X is stochastically less than or equal to Y, written $X \leq_{\text{st}} Y$ if*

$$P(X \leq w) \geq P(Y \leq w) \quad \text{for all } w.$$

The following lemma states that the objective function of a point sampled from a Boltzmann distribution with temperature T is stochastically less than or equal to the objective function of a point sampled from a Boltzmann distribution with temperature T' for all $T \leq T'$. This was noted by Romeijn and Smith [139], and is used in the performance analysis in Section 2.

LEMMA 4.1 *Suppose X has a Boltzmann distribution with parameter T and density $g_T(x)$, and X' has a Boltzmann distribution with parameter T' and density $g_{T'}(x)$, where $0 < T \leq T'$. Let $Y = f(X)$ and $Y' = f(X')$. Then,*

$$Y \leq_{\text{st}} Y',$$

that is,

$$P(f(X) \leq y) \geq P(f(X') \leq y) \quad \text{for all } y.$$

Proof: We want to show that $P(f(X) \leq y) \geq P(f(X') \leq y)$ for all y and $0 < T \leq T'$. We write

$$P(f(X) \leq y) = \int_{x:f(x)\leq y} g_T(x)dx = \int_{x\in S(y)} \frac{e^{-f(x)/T}dx}{\int_{z\in S} e^{-f(z)/T}dz}.$$

Define the nonnegative quantities

$$\begin{aligned} a(y,T) &= \int_{x:f(x)\leq y} e^{\left(\frac{y-f(x)}{T}\right)}\,dx \\ b(y,T) &= \int_{x:f(x)> y} e^{\left(\frac{y-f(x)}{T}\right)}\,dx \end{aligned}$$

so that, with $X \sim g_T$, we can write

$$P(f(X) \leq y) = \frac{a(y,T)}{a(y,T)+b(y,T)}.$$

Now, taking the derivative with respect to T,

$$\begin{aligned} \frac{d}{dT}a(y,T) &= \int_{x:f(x)\leq y} \frac{d}{dT} e^{\left(\frac{y-f(x)}{T}\right)}\,dx \\ &= \int_{x:f(x)\leq y} \left(\frac{f(x)-y}{T^2}\right) e^{\left(\frac{y-f(x)}{T}\right)}\,dx \\ &\leq 0. \end{aligned}$$

Similarly, $db(y,T)/dT \geq 0$. Noting that

$$\frac{d}{dT}P(f(X) \leq y) = (a/(a+b))' = (ba' - ab')/(a+b)^2,$$

it follows that

$$\frac{d}{dT}P(f(X) \leq y) \leq 0$$

implying $P(f(X) \leq y)$ is nonincreasing in T and proving the result. ■

The dominance property of the Boltzmann distribution provides some insight into why the Boltzmann distribution is a good choice for simulated annealing. If a simulated annealing algorithm could sample exactly from a Boltzmann distribution, then for small T values, the distribution has a high likelihood of sampling close to the global optimum. Then, one might ask, why doesn't one choose $T = 0$ from the beginning? The answer is that it is impractical to implement, because the Boltzmann distribution with $T = 0$ is just a degenerate distribution with all probability mass at the global optimum. The hope is that a bootstrapping approach to approximating the Boltzmann distribution as T decreases will be more efficient than pure random search, and may be closer to the ideal performance of pure adaptive search. The temperature values are updated by means of the cooling schedule, $T_{k+1} = \tau(Y_{k+1})$.

We are now ready to define the annealing adaptive search algorithm as follows.

Annealing Adaptive Search (AAS)

Step 0. Initialize $X_0 \in S$ according to a uniform distribution (i.e., Boltzmann distribution on S with parameter $T = \infty$). Set $k = 0$. Set $Y_0 = f(X_0)$ and set $T_0 = \tau(Y_0)$ where τ is a non-negative real-valued nonincreasing function.

Step 1. Generate X_{k+1} according to a Boltzmann distribution on S with parameter T_k. Set

$$Y_{k+1} = \begin{cases} f(X_{k+1}) & \text{if } f(X_{k+1}) < Y_k \\ Y_k & \text{otherwise} \end{cases}$$

and set $X_{\text{best}} = X_{k+1}$ if $f(X_{k+1}) < Y_k$. Update $T_{k+1} = \tau(Y_{k+1})$.

Step 2. If a stopping criterion is met, stop. Otherwise, increment k and return to Step 1.

Annealing adaptive search is an abstracted form of simulated annealing, which samples from the original feasible region S using the Boltzmann distribution to focus the sample points around the global minima. It is still not readily implemented because it assumes the samples are drawn exactly from a Boltzmann distribution, while realistic algorithms may only approximate a Boltzmann distribution. AAS is based on the adaptive search (AS) algorithm introduced by Romeijn and Smith [139]. The difference between AAS and AS is that AAS explicitly counts all sample points while AS counts only the points that are improving.

The adaptive search algorithm, from [139], follows.

Adaptive Search (AS)

Step 0. Initialize X_0 according to a Boltzmann distribution on S with parameter $T = \infty$. Set $k = 0$. Set $Y_0 = f(X_0)$ and set $T_0 = \tau(Y_0)$ where τ is a non-negative real-valued nonincreasing function.

Step 1. Generate X_{k+1} according to a Boltzmann distribution on S with parameter T_k, repeatedly and independently, until improvement $f(X_{k+1}) < f(X_k)$. Set $Y_{k+1} = f(X_{k+1})$. Update $T_{k+1} = \tau(Y_{k+1})$.

Step 2. If a stopping criterion is met, stop. Otherwise increment k and return to Step 1.

The temperature parameter T in both AAS and AS depends on the best objective function value sampled, and is modified only when improving points are found, $T_{k+1} = \tau(Y_{k+1})$, because Y_{k+1} only changes value when the sample point is improving. It is assumed that the cooling schedule τ is a nonincreasing function, however one could imagine the algorithm if the cooling schedule was fixed at infinity, i.e. $\tau(y) \equiv \infty$. Then AAS becomes PRS, since the Boltzmann distribution with parameter infinity is equivalent to the uniform distribution, and AS becomes PAS. Thus, intuitively, AAS will converge to the global optimum at least as fast as PRS, and AS will converge at least as fast as PAS, since each iteration has a temperature parameter less than ∞. These ideas are formalized in the next section, where the effect of reducing the temperature to zero can be seen to improve convergence rate. Analytically derived cooling schedules are discussed in Section 3.

We now turn to deriving bounds on the number of improving points (record values) and the number of sample points (iterations) in annealing adaptive search.

2. Bounds on Performance of Annealing Adaptive Search

The performance analysis of annealing adaptive search relies on previous results for PAS and HAS, as well as results developed by Romeijn and Smith [139] on adaptive search. We first discuss the relationship between AAS and adaptive search.

Annealing adaptive search is related to adaptive search through its record values. In Lemma 2.1, we showed that the record values of PRS are equal in distribution to the iterates of PAS. Analogously, we show in Lemma 4.2 that the record values of AAS are equal in distribution to the iterates of AS.

Let Y_k^{AAS} denote the objective function value on iteration k of AAS, and following the notation in Chapter 2, let $Y_{R(k)}^{\text{AAS}}$ be the objective function value of the kth record AAS. Similarly, let Y_k^{AS} denote the objective function value on iteration k of adaptive search.

LEMMA 4.2 *For a general global optimization problem* (P), *the stochastic process for adaptive search* $\{Y_k^{\text{AS}},\ k = 0,1,2,\ldots\}$ *is equal in distribution to the process* $\left\{Y_{R(k)}^{\text{AAS}},\ k = 0,1,2,\ldots\right\}$ *where* $R(k)$ *is the* k^{th} *record of the AAS sequence* $\left\{Y_k^{\text{AAS}},\ k = 0,1,2,\ldots\right\}$, *that is,*

$$\left\{Y_k^{\text{AS}}\right\} \sim \left\{Y_{R(k)}^{\text{AAS}}\right\}$$

and specifically,

$$P\left(Y_k^{\text{AS}} \leq y\right) = P\left(Y_{R(k)}^{\text{AAS}} \leq y\right) \quad \textit{for } k = 0,1,2,\ldots \textit{ and } y_* \leq y \leq y^*.$$

Proof: First, we show that the conditional distributions are equal,

$$P\left(Y_{R(k+1)}^{\text{AAS}} \leq z | Y_{R(k)}^{\text{AAS}} = y\right) = P\left(Y_{k+1}^{\text{AS}} \leq z | Y_k^{\text{AS}} = y\right).$$

for all $k = 0,1,2,\ldots$ *and* $y_* \leq y \leq y^*$.

Let k be any iteration, $k = 0,1,2,\ldots$. If $z \geq y$ then both conditional probabilities equal one, so consider z, y such that $y_* \leq z < y \leq y^*$. Then,

$$\begin{aligned}
&P\left(Y_{R(k+1)}^{\text{AAS}} \leq z | Y_{R(k)}^{\text{AAS}} = y\right) \\
&= P\left(Y_{R(k)+1}^{\text{AAS}} \leq z | Y_{R(k)}^{\text{AAS}} = y\right) \\
&\quad + P\left(Y_{R(k)+2}^{\text{AAS}} \leq z, Y_{R(k)+1}^{\text{AAS}} \geq y | Y_{R(k)}^{\text{AAS}} = y\right) \\
&\quad + P\left(Y_{R(k)+3}^{\text{AAS}} \leq z, Y_{R(k)+2}^{\text{AAS}} \geq y, Y_{R(k)+1}^{\text{AAS}} \geq y | Y_{R(k)}^{\text{AAS}} = y\right) + \cdots
\end{aligned}$$

and because each AAS iteration is independent and identically distributed with Boltzmann $T = \tau(y)$,

$$
\begin{aligned}
&= P\left(Y^{\text{AAS}}_{R(k)+1} \le z | Y^{\text{AAS}}_{R(k)} = y\right) \\
&\quad + P\left(Y^{\text{AAS}}_{R(k)+1} \le z | Y^{\text{AAS}}_{R(k)} = y\right) \cdot P\left(Y^{\text{AAS}}_{R(k)+1} \ge y | Y^{\text{AAS}}_{R(k)} = y\right) \\
&\quad + P\left(Y^{\text{AAS}}_{R(k)+1} \le z | Y^{\text{AAS}}_{R(k)} = y\right) \cdot P\left(Y^{\text{AAS}}_{R(k)+1} \ge y | Y^{\text{AAS}}_{R(k)} = y\right)^2 + \cdots \\
&= P\left(Y^{\text{AAS}}_{R(k)+1} \le z | Y^{\text{AAS}}_{R(k)} = y\right) \sum_{i=0}^{\infty} P\left(Y^{\text{AAS}}_{R(k)+1} \ge y | Y^{\text{AAS}}_{R(k)} = y\right)^i \\
&= P\left(Y^{\text{AAS}}_{R(k)+1} \le z | Y^{\text{AAS}}_{R(k)} = y\right) \cdot \frac{1}{1 - P\left(Y^{\text{AAS}}_{R(k)+1} \ge y | Y^{\text{AAS}}_{R(k)} = y\right)} \\
&= \frac{\int_{x \in S(z)} e^{-f(x)/\tau(y)} dx}{\int_{x \in S} e^{-f(x)/\tau(y)} dx} \cdot \frac{1}{1 - P\left(Y^{\text{AAS}}_{R(k)+1} \ge y | Y^{\text{AAS}}_{R(k)} = y\right)} \\
&= \frac{\int_{x \in S(z)} e^{-f(x)/\tau(y)} dx}{\int_{x \in S} e^{-f(x)/\tau(y)} dx} \cdot \frac{\int_{x \in S} e^{-f(x)/\tau(y)} dx}{\int_{x \in S(y)} e^{-f(x)/\tau(y)} dx} \\
&= \frac{\int_{x \in S(z)} e^{-f(x)/\tau(y)} dx}{\int_{x \in S(y)} e^{-f(x)/\tau(y)} dx} \\
&= \frac{\pi_{\tau(y)}(S(z))}{\pi_{\tau(y)}(S(y))},
\end{aligned}
$$

where $\pi_{\tau(y)}$ is the Boltzmann distribution with parameter $\tau(y)$, which is the conditional distribution of improvement for adaptive search,

$$
= P\left(Y^{\text{AS}}_{k+1} \le z | Y^{\text{AS}}_{k} = y\right).
$$

Thus, we have

$$
P\left(Y^{\text{AAS}}_{R(k+1)} \le z | Y^{\text{AAS}}_{R(k)} = y\right) = P\left(Y^{\text{AS}}_{k+1} \le z | Y^{\text{AS}}_{k} = y\right).
$$

Next, induction is used to show that the unconditional distributions are equal, as in Lemma 2.1. For $k = 0$, one has $T = +\infty$ and thus has a uniformly distributed point over S for either AAS or AS. Therefore,

$$
\begin{aligned}
P\left(Y^{\text{AS}}_0 \le y\right) &= P\left(Y^{\text{AAS}}_0 \le y\right), \\
&= P\left(Y^{\text{AAS}}_{R(0)} \le y\right),
\end{aligned}
$$

and hence,

$$
Y^{\text{AAS}}_{R(0)} \sim Y^{\text{AS}}_0.
$$

Let k be an integer greater than 0, and suppose $Y_{R(i)}^{\text{AAS}} \sim Y_i^{\text{AS}}$ for $i = 0, 1, 2, \ldots, k$. Then,

$$P\left(Y_{R(k+1)}^{\text{AAS}} \leq y\right) = E\left[P\left(Y_{R(k+1)}^{\text{AAS}} \leq y | Y_{R(k)}^{\text{AAS}}\right)\right]$$

and using the equality of the conditional distributions as well as the induction hypothesis of equality of the marginal distributions,

$$= E\left[P\left(Y_{k+1}^{\text{AS}} \leq y | Y_k^{\text{AS}}\right)\right]$$

$$= P\left(Y_{k+1}^{\text{AS}} \leq y\right) \qquad \text{for all } y_* \leq y \leq y^*.$$

Therefore,

$$Y_{R(k+1)}^{\text{AAS}} \sim Y_{k+1}^{\text{AS}}$$

and by induction the two sequences are equal in distribution. ∎

Not only is annealing adaptive search related to adaptive search, it is also related to hesitant adaptive search. To analyze the performance of AAS, we are not only interested in the number of record values, but also in the number of sample points needed to obtain a record value. We can use HAS to analyze the number of sample points in AAS by choosing the underlying sampling distribution δ, and the "bettering" probability $b(y)$, appropriately.

Consider HAS with uniform sampling (δ is uniform), and let the probability of sampling an improving point, or bettering probability be,

$$b(y) = \frac{\int_{x \in S(y)} e^{-f(x)/\tau(y)} dx}{\int_{x \in S} e^{-f(x)/\tau(y)} dx} \tag{4.2}$$

for $y_* \leq y \leq y^*$. Let this special case of HAS be denoted HAS′. The expression for $b(y)$ can be interpreted as the probability of generating an improving point, given a current objective function value of y, using a Boltzmann distribution over S with parameter $\tau(y)$. This bettering probability combined with a uniform distribution provides the relationship between AAS and HAS. Let $\{Y_k^{\text{HAS}'}, k = 0, 1, 2, \ldots\}$ be the objective function values for HAS′, that is, with uniform sampling and $b(y)$ as in Equation 4.2. The lemma below states that the objective function values of AAS are stochastically less than or equal to those of HAS′ implying that AAS performs at least as well as this special case of HAS.

THEOREM 4.3 *For a global optimization problem (P), the stochastic process for annealing adaptive search is stochastically less than or equal to the process of hesitant adaptive search with a uniform distribution and*

$$b(y) = \frac{\int_{x \in S(y)} e^{-f(x)/\tau(y)} dx}{\int_{x \in S} e^{-f(x)/\tau(y)} dx}$$

that is,

$$\left\{Y_k^{\mathrm{AAS}}\right\} \leq_{\mathrm{st}} \left\{Y_k^{\mathrm{HAS}'}\right\}$$

and specifically,

$$P\left(Y_k^{\mathrm{AAS}} \leq y\right) \geq P\left(Y_k^{\mathrm{HAS}'} \leq y\right) \quad \textit{for } k = 0, 1, 2, \ldots \textit{ and } y_* \leq y \leq y^*.$$

Proof: We first show that the conditional probability of AAS $P(Y_{k+1}^{\mathrm{AAS}} \leq y | Y_k^{\mathrm{AAS}} = z)$ is nonincreasing in z for any fixed k, $k = 0, 1, 2, \ldots$ and y values, $y_* \leq y \leq y^*$. Let z and z' be any two real valued variables satisfying $z < z'$. When $y > z$, then

$$P\left(Y_{k+1}^{\mathrm{AAS}} \leq y | Y_k^{\mathrm{AAS}} = z\right) = 1 \geq P\left(Y_{k+1}^{\mathrm{AAS}} \leq y | Y_k^{\mathrm{AAS}} = z'\right).$$

Therefore the conditional probability of AAS is nonincreasing when the values of y and z satisfy $y > z$. Now consider y, z such that $y \leq z$. Then, for any k, $k = 0, 1, 2, \ldots$, one has,

$$P\left(Y_{k+1}^{\mathrm{AAS}} \leq y | Y_k^{\mathrm{AAS}} = z\right) = \frac{\int_{x \in S(y)} e^{-f(x)/\tau(z)} dx}{\int_{x \in S} e^{-f(x)/\tau(z)} dx}$$

and apply Lemma 4.1 to the last term, viewing $T = \tau(z)$ and $T' = \tau(z')$ and using the assumption that $\tau(u)$ is nonincreasing in u, we have

$$\begin{aligned} &\geq \frac{\int_{x \in S(y)} e^{-f(x)/\tau(z')} dx}{\int_{x \in S} e^{-f(x)/\tau(z')} dx} \\ &= P\left(Y_{k+1}^{\mathrm{AAS}} \leq y | Y_k^{\mathrm{AAS}} = z'\right). \end{aligned}$$

Therefore, for all fixed y and k, $k = 0, 1, 2, \ldots$, $P\left(Y_{k+1}^{\mathrm{AAS}} \leq y | Y_k^{\mathrm{AAS}} = z\right)$ is nonincreasing in z.

Next, we compare the conditional probabilities for AAS and HAS′, and show that $P(Y_{k+1}^{\mathrm{AAS}} \leq y | Y_k^{\mathrm{AAS}} = z) \geq P(Y_{k+1}^{\mathrm{HAS}'} \leq y | Y_k^{\mathrm{HAS}'} = z)$ for all $k = 0, 1, 2, \ldots$ and $y_* \leq y \leq y^*$. If $y \geq z$, the conditional probabilities are both equal to 1, and so satisfy the inequality. Now consider y, z such that $y < z$. Then, for any k, $k = 0, 1, 2, \ldots$, the conditional probability of AAS is,

$$P\left(Y_{k+1}^{\mathrm{AAS}} \leq y | Y_k^{\mathrm{AAS}} = z\right) =$$

$$= \frac{\int_{x\in S(y)} e^{-f(x)/\tau(z)}dx}{\int_{x\in S} e^{-f(x)/\tau(z)}dx}$$

$$= \frac{\int_{x\in S(y)} e^{-f(x)/\tau(z)}dx}{\int_{x\in S} e^{-f(x)/\tau(z)}dx} \cdot \frac{\int_{x\in S(z)} e^{-f(x)/\tau(z)}dx}{\int_{x\in S(z)} e^{-f(x)/\tau(z)}dx}$$

$$= b(z)\frac{\int_{x\in S(y)} e^{-f(x)/\tau(z)}dx}{\int_{x\in S(z)} e^{-f(x)/\tau(z)}dx}$$

and apply Lemma 4.1 viewing $T = \tau(z)$ and $T' = +\infty$, yielding

$$\geq b(z)\frac{\int_{x\in S(y)} e^{-f(x)/+\infty}dx}{\int_{x\in S(z)} e^{-f(x)/+\infty}dx},$$

$$= b(z)\frac{\nu(S(y))}{\nu(S(z))}$$

where ν denotes Lebesgue measure, and because HAS$'$ uses the uniform distribution,

$$= P\left(Y_{k+1}^{\text{HAS}'} \leq y | Y_k^{\text{HAS}'} = z\right).$$

Finally we show that the sequence $\{Y_k^{\text{AAS}}\}$ is stochastically less than or equal to the sequence $\{Y_k^{\text{HAS}'}\}$ by using induction. For $k = 0$, both AAS and HAS$'$ sample according to a uniform distribution over S, hence

$$\left\{Y_0^{\text{AAS}}\right\} \leq_{\text{st}} \left\{Y_0^{\text{HAS}}\right\}.$$

Let k be an integer greater than 0, and suppose $\left\{Y_i^{\text{AAS}}\right\} \leq_{\text{st}} \left\{Y_i^{\text{HAS}}\right\}$, i.e., $P\left(Y_i^{\text{AAS}} \leq y\right) \geq P\left(Y_i^{\text{HAS}'} \leq y\right)$, for $i = 1, 2, \ldots, k$. Then for any $y_* \leq y \leq y^*$,

$$\begin{aligned} P\left(Y_{k+1}^{\text{AAS}} \leq y\right) &= E\left[P\left(Y_{k+1}^{\text{AAS}} \leq y | Y_k^{\text{AAS}}\right)\right] \\ &= \int_{z=y_*}^{z=y^*} P\left(Y_{k+1}^{\text{AAS}} \leq y | Y_k^{\text{AAS}} = z\right) dP\left(Y_k^{\text{AAS}} \leq z\right) \end{aligned}$$

and letting $h(z) = P\left(Y_{k+1}^{\text{AAS}} \leq y | Y_k^{\text{AAS}} = z\right)$, we have

$$= \int_{z=y_*}^{z=y^*} h(z) dP\left(Y_k^{\text{AAS}} \leq z\right)$$

and because $h(y_*) = 1$ and $h(z)$ is differentiable, we have

$$= \int_{z=y_*}^{z=y^*} \left(1 + \int_{u=y_*}^{u=z} h'(u)du\right) dP\left(Y_k^{\text{AAS}} \leq z\right)$$

$$= \; 1+\int_{z=y_*}^{z=y^*}\left(\int_{u=y_*}^{u=z} h'(u)du\right)dP\left(Y_k^{\text{AAS}} \le z\right).$$

Switching the order of the integrations, we have

$$= \; 1+\int_{u=y_*}^{u=y^*} h'(u)\left(\int_{z=u}^{z=y^*} dP\left(Y_k^{\text{AAS}} \le z\right)\right)du$$

$$= \; 1+\int_{u=y_*}^{u=y^*} h'(u)\left(1-P\left(Y_k^{\text{AAS}} \le u\right)\right)du$$

and because $h(u)$ is nonincreasing, which implies $h'(u) \le 0$, combined with the induction assumption, we have

$$\ge \; 1+\int_{u=y_*}^{u=y^*} h'(u)\left(1-P\left(Y_k^{\text{HAS}'} \le u\right)\right)du$$

which is equivalent to

$$= \; 1+\int_{u=y_*}^{u=y^*} h'(u)\left(\int_{z=u}^{z=y^*} dP\left(Y_k^{\text{HAS}'} \le z\right)\right)du.$$

Switching the order of the integrations back, we have

$$\begin{aligned}
&= \; 1+\int_{z=y_*}^{z=y^*}\left(\int_{u=y_*}^{u=z} h'(u)du\right)dP\left(Y_k^{\text{HAS}'} \le z\right) \\
&= \; 1+\int_{z=y_*}^{z=y^*} (h(z)-1)\,dP\left(Y_k^{\text{HAS}'} \le z\right) \\
&= \; \int_{z=y_*}^{z=y^*} h(z)dP\left(Y_k^{\text{HAS}'} \le z\right) \\
&= \; \int_{z=y_*}^{z=y^*} P\left(Y_{k+1}^{\text{AAS}} \le y | Y_k^{\text{AAS}} = z\right)dP\left(Y_k^{\text{HAS}'} \le z\right)
\end{aligned}$$

and since we already showed that $P(Y_{k+1}^{\text{AAS}} \le y|Y_k^{\text{AAS}} = z) \ge P(Y_{k+1}^{\text{HAS}'} \le y|Y_k^{\text{HAS}'} = z)$, we get

$$\begin{aligned}
&\ge \; \int_{z=y_*}^{z=y^*} P\left(Y_{k+1}^{\text{HAS}'} \le y | Y_k^{\text{HAS}'} = z\right)dP\left(Y_k^{\text{HAS}'} \le z\right) \\
&= \; P\left(Y_{k+1}^{\text{HAS}'} \le y\right).
\end{aligned}$$

So we conclude that $P\left(Y_{k+1}^{\text{AAS}} \le y\right) \ge P\left(Y_{k+1}^{\text{HAS}'} \le y\right)$. ■

Before summarizing the relationships developed thus far, recall that W_k^{PAS} denotes the objective function value on iteration k of pure adaptive search. Theorem 4.3 states that the objective function values of

AAS are stochastically less than or equal to those of HAS with uniform sampling and $b(y)$ defined in Equation 4.2,

$$\left\{Y_k^{AAS}\right\} \leq_{\text{st}} \left\{Y_k^{HAS'}\right\}$$

while Lemma 4.2 states that the record values of AAS are equal in distribution to the iterates of AS,

$$\left\{Y_{R(k)}^{AAS}\right\} \sim \left\{Y_k^{AS}\right\}.$$

In addition, the record values of HAS′ are equal in distribution to the iterates of PAS,

$$\left\{Y_{R(k)}^{HAS'}\right\} \sim \left\{W_k^{PAS}\right\},$$

because the record values of HAS′ depend only on the underlying sampling distribution, which is uniform, and not on the bettering probability $b(y)$. Consequently, if the objective function values of AAS are stochastically less than or equal to those of HAS′, then intuitively, the record values of AAS are stochastically less than or equal to those of HAS′, that is PAS. Thus, the performance of AS is better than that of PAS. This is formalized in the following theorem, and is based on the result by Romeijn and Smith [139, Theorem 3.5].

THEOREM 4.4 *Given AAS, AS, PAS and HAS′, which uses uniform sampling and $b(y)$ as in Equation 4.2 on a global optimization problem (P), then*

$$\left\{Y_{R(k)}^{AAS}\right\} \sim \left\{Y_k^{\text{AS}}\right\} \leq_{\text{st}} \left\{W_k^{\text{PAS}}\right\} \sim \left\{Y_{R(k)}^{HAS'}\right\}$$

for $k = 0, 1, \ldots$.

Proof: Lemma 4.2 gives us that $\left\{Y_{R(k)}^{AAS}\right\} \sim \left\{Y_k^{\text{AS}}\right\}$, and $\left\{W_k^{\text{PAS}}\right\} \sim \left\{Y_{R(k)}^{HAS'}\right\}$ because of the choice of uniform distribution for HAS′, so it remains to show that $\left\{Y_k^{\text{AS}}\right\} \leq_{\text{st}} \left\{W_k^{\text{PAS}}\right\}$. Using the same approach as in Lemma 4.2 and Theorem 4.3, we start by showing that the conditional distributions are stochastically dominated. For all $k = 0, 1, 2, \ldots$, and $y_* \leq z \leq y \leq y^*$, the conditional probability of AS is,

$$P\left(Y_{k+1}^{\text{AS}} < z | Y_k^{\text{AS}} = y\right) = \frac{\int_{x\in S(z)} e^{-f(x)/\tau(y)}dx}{\int_{x\in S(y)} e^{-f(x)/\tau(y)}dx}$$

and using Lemma 4.1 with $T = \tau(y)$ and $T' = +\infty$,

$$\begin{aligned} &\geq \frac{\int_{x\in S(z)} e^{-f(x)/+\infty} dx}{\int_{x\in S(y)} e^{-f(x)/+\infty} dx} \\ &= \frac{\nu(S(z))}{\nu(S(y))} \end{aligned}$$

where ν denotes uniform distribution,

$$= P\left(W_{k+1}^{\text{PAS}} < z | W_k^{\text{PAS}} = y\right).$$

Next, induction is used to uncondition and show that the sequence $\left\{Y_k^{\text{AS}}\right\}$ is stochastically less than or equal to the sequence $\left\{W_k^{\text{PAS}}\right\}$. For $k = 0$, AS sets $T = +\infty$ which implies Y_0^{AS} is uniformly distributed over S, and the initial point in PAS is also sampled uniformly over S, hence,

$$Y_0^{\text{AS}} \leq_{\text{st}} W_0^{\text{PAS}}.$$

Now let k be an integer greater than 0, and suppose $Y_i^{\text{AS}} \leq_{\text{st}} W_i^{\text{PAS}}$ for $i = 0, 1, 2, \ldots, k$. For any $y_* \leq y \leq y^*$,

$$P\left(Y_{k+1}^{\text{AS}} \leq y\right) = E\left[P\left(Y_{k+1}^{\text{AS}} \leq y | Y_k^{\text{AS}}\right)\right]$$

and following the argument in Theorem 4.3, because the conditional probabilities are dominated and $P\left(Y_k^{\text{AS}} < y | Y_k^{\text{AS}} = z\right)$ is nonincreasing in z due to $\tau(z)$ nonincreasing, and because the marginal probabilities are dominated by the induction hypothesis, we have

$$\begin{aligned} &\geq E\left[P\left(W_{k+1}^{\text{PAS}} \leq y | W_k^{\text{PAS}}\right)\right] \\ &= P\left(W_{k+1}^{\text{PAS}} \leq y\right). \end{aligned}$$

Thus, $P\left(Y_{k+1}^{\text{AS}} \leq y\right) \geq P\left(W_{k+1}^{\text{PAS}} \leq y\right)$, which proves the theorem. ■

The measure of performance of annealing adaptive search is stated in terms of the number of iterations needed to first achieve y or better. Let $N^{\text{AAS}}(y)$, $N^{\text{AS}}(y)$, $N^{\text{PAS}}(y)$ and $N^{\text{HAS}'}(y)$ denote the number of iterations for AAS, AS, PAS and HAS′, respectively, required to achieve a value of y or better, for $y_* < y \leq y^*$.

The following corollary gives an upper bound on the number of sample points (e.g. function evaluations) of annealing adaptive search as well as an upper bound on the number of improving points (e.g. record values) of annealing adaptive search to achieve a level of performance, in terms of HAS′ and PAS, respectively.

COROLLARY 4.5 *For a general global optimization problems (P), the following relationships hold for the number of iterations required to achieve a value of y or better,*

$$\begin{aligned} N^{\mathrm{AAS}}(y) &\leq_{\mathrm{st}} N^{\mathrm{HAS}'}(y), \textit{ and} \\ N^{\mathrm{AS}}(y) &\leq_{\mathrm{st}} N^{\mathrm{PAS}}(y) \end{aligned}$$

for $y_* \leq y \leq y^*$.

Proof: The two statements are immediate from Theorems 4.3 and 4.4 and noting that

$$Y_k < y \text{ if and only if } N(y) < k+1$$

so $P(Y_k < y) = P(N(y) < k+1)$ for $k = 0, 1, \ldots$, as in Chapter 2. ∎

The distributions for $N(y)$ are known for PAS and HAS′, as described in Chapters 2 and 3. In particular, the bounds on the expected number of iterations is given in the following corollaries. Corollary 4.6 gives an upper bound on the expected number of function evaluations of annealing adaptive search and Corollary 4.7 gives an upper bound on the expected number of record values (improving points) of annealing adaptive search to achieve a level of performance. The expected number of record values of AAS is bounded by PAS, and therefore is linear in the dimension of the problem.

COROLLARY 4.6 *For a general global optimization (P), the expected number of AAS iterations is bounded above by the expected number of HAS′, iterations, which uses uniform sampling and b(y) as in Equation 4.2,*

$$\begin{aligned} E[N^{\mathrm{AAS}}(y)] &\leq E[N^{\mathrm{HAS}'}(y)] \\ &\leq 1 + \int_{y^+}^{\infty} \frac{d\rho(t)}{b(t)p(t)} \end{aligned}$$

for $y_* \leq y \leq y^*$.

Proof: The result is immediate from Corollary 4.5 and Corollary 3.7. ∎

COROLLARY 4.7 *If the global optimization problem (P) is over a convex feasible region in n dimensions with diameter at most D, and with Lipschitz constant at most K, the expected number of record values of AAS required to achieve a value of y of better, is equal to* $E[N^{\text{AS}}(y)]$ *with an upper bound of,*

$$E[N^{\text{AS}}(y)] \;\leq\; 1 + n\ln(KD/\,(y - y_*))$$

for $y_* \leq y \leq y^*$.

Similarly, for finite global optimization problems (P) *with at most* K^n *distinct objective function values,*

$$E[N^{\text{AS}}(y_*)] \;\leq\; 1 + n\ln(K).$$

Proof: The result is immediate from Corollary 4.6 and Corollaries 2.10 and 2.12. ■

Hesitant adaptive search now provides an upper bound on the total number of sample points needed by annealing adaptive search to achieve an objective function value of y or better, and pure adaptive search provides the linearity bound on the number of improving points needed. This provides some insights into why simulated annealing is reported to have good performance. If the algorithm is able to generate improving points close to a Boltzmann distribution, then it is expected it will perform well. If the algorithm gets stuck in a local optimum, then it is not approximating the Boltzmann distribution well at all, and is resorting to pure random search. This understanding of the behavior of the adaptive annealing search is next used to motivate an analytical cooling schedule.

3. Cooling Schedule for Annealing Adaptive Search

The cooling schedule for annealing adaptive search, $T_k = \tau(Y_k)$, affects the number of improving points required, as well as the number of sample points needed to achieve a record value. There is a tradeoff in computational effort. If T_k decreases rapidly, the number of record values is small, while the number of sample points needed to achieve a record value is large. Conversely, if T_k is reduced slowly, the number of record values is large, while the number of sample points needed to achieve a record value is hopefully small.

Two cooling schedules have been derived in the context of continuous domains, that keep the number of sample points needed to generate an improving point small, and solve for the temperature parameter T based on the form of the optimization problem. Romeijn and Smith [139, 140] derived a cooling schedule based on a quadratic approximation to the objective function. Kiatsupaibul and Smith [85, 86] generalized this cooling schedule, assuming a Lipschitz continuous function.

The idea of Romeijn and Smith [139, 140] is to reduce the temperature of the Boltzmann distribution in such a way that the probability of generating an improving point remains high. Specifically, given a current objective function value Y_k, and using a Boltzmann distribution with temperature T_{k+1}, find $T_{k+1} = \tau(Y_k)$ such that,

$$P(f(X_{k+1}) < Y_k) \geq 1 - \alpha,$$

where $0 < \alpha < 1$. For example, if $\alpha = 0.05$, then the probability of finding an improving point on a single sample would be $1 - \alpha = 0.95$ and the expected number of samples to obtain an improving point, $1/(1-\alpha)$, is slightly over one.

The following theorem is based on [140], and provides the value of T using a chi-squared distribution, for a strictly convex quadratic objective function where the feasible region is a level set of f. Although the intent is to apply the cooling schedule to a nonconvex function, one approach is to approximate f using a second-order Taylor series expansion about a point $x_0 \in S$, where

$$f(x) \approx (x - x_0)'H(x - x_0) - 2r'(x - x_0) + f(x_0)$$

and H is a positive definite $n \times n$ matrix and r is an $n \times 1$ vector. For convenience, in Theorem 4.8 we take $x_0 = 0$ and $f(x_0) = 0$ without loss of generality, as in [140].

THEOREM 4.8 *For $f(x) = x'Hx - 2r'x$ where H is a positive definite $n \times n$ matrix and r is an $n \times 1$ vector, and the feasible region is a level set of f, i.e., $S = \{x \in \mathbb{R}^n : f(x) \leq y_* + c\}$ for $c > 0$, then if X is generated from a Boltzmann distribution on S with parameter $T = \tau(y)$ for $y_* < y \leq y_* + c$, and $\alpha \in (0, 1)$ we have that*

$$P(f(X) < y) \geq 1 - \alpha$$

if

$$T = \tau(y) = \frac{2(y - y_*)}{\chi^2_{1-\alpha}(n)}$$

where $\chi^2_{1-\alpha}(n)$ is the $100(1 - \alpha)$ percentile point of the chi-squared distribution with n degrees of freedom.

Proof: The proof follows Romeijn and Smith [140] pages 111-112. Let X be a random variable with distribution π_T on S, then for $y_* < y \leq y_* + c$,

$$P(f(X) < y) \quad = \quad \frac{\int_{x \in S(y)} e^{-f(x)/T} dx}{\int_{x \in S} e^{-f(x)/T} dx}$$

$$\geq \frac{\int_{x\in S(y)} e^{-f(x)/T} dx}{\int_{x\in\mathbb{R}^n} e^{-f(x)/T} dx}. \quad (4.3)$$

Letting $u = \sqrt{\frac{2}{T}} H^{1/2}(x - H^{-1}r)$, one can verify that

$$u'u/2 = (f(x) - y_*)/T$$

and that the set $S(y) = \{x \in \mathbb{R}^n : f(x) < y\}$ is equivalent to the set $\hat{S}(y) = \{u \in \mathbb{R}^n : u'u < 2(y - y_*)/T\}$. Consequently, transforming the numerator and denominator of Equation 4.3 with $u = \sqrt{\frac{2}{T}} H^{1/2}(x - H^{-1}r)$, we get

$$P(f(X) < y) \geq \frac{\int_{u\in\hat{S}(y)} e^{-u'u/2} du}{\int_{u\in\mathbb{R}^n} e^{-u'u/2} du}.$$

Also, if $||U||^2$ has a chi-squared distribution with n degrees of freedom,

$$P(||U||^2 \leq 2(y - y_*)/T) = \frac{\int_{u\in\hat{S}(y)} e^{-u'u/2} dx}{\int_{u\in\mathbb{R}^n} e^{-u'u/2} dx}.$$

Hence, this probability equals $1 - \alpha$, if

$$2(y - y_*)/T = \chi^2_{1-\alpha}(n)$$

or equivalently, if

$$T = \frac{2(y - y_*)}{\chi^2_{1-\alpha}(n)}$$

where $\chi^2_{1-\alpha}(n)$ satisfies $P(||U||^2) \leq \chi^2_{1-\alpha}(n)) = 1 - \alpha$. Consequently, we have that

$$P(f(X) < y) \geq 1 - \alpha \text{ if } T = \frac{2(y - y_*)}{\chi^2_{1-\alpha}(n)}$$

where $\chi^2_{1-\alpha}(n)$ is the $100(1 - \alpha)$ percentile point of the chi-squared distribution with n degrees of freedom. ∎

In Romeijn and Smith [139], a slightly different cooling schedule for adaptive search is given, which relies on the gamma distribution. Three examples include:

Example 1. Linear Programming: if f is linear and S is a full n-dimensional polytope, and the optimal solution is unique and nondegenerate, then the cooling schedule is

$$\tau(y) = \frac{y - y_*}{\gamma_{1-\alpha}(n, 1)},$$

Example 2. Concave Minimization: if f is concave and S is a full n-dimensional polytope, then the cooling schedule is the same as in Example 1,

Example 3. Convex Minimization: if f is a strictly convex quadratic function and S is a level set of f, then the cooling schedule is

$$\tau(y) = \frac{y - y_*}{\gamma_{1-\alpha}(n/2, 1)},$$

where $\gamma_{1-\alpha}(n, 1)$ satisfies

$$P(V \leq \gamma_{1-\alpha}(n, 1)) = 1 - \alpha$$

and V has a gamma distribution with parameters n and 1. For the full details of their analysis, see pages 583-585 in [139].

In the cooling schedules given in Theorem 4.8 and the above examples, the value of the global minimum y_* is needed. To implement the cooling schedule, we can replace y_* with any consistent estimator $\hat{f}(x)$ of the global minimum based on the objective function values observed during the course of the algorithm. A consistent estimator $\hat{f}(x)$, suggested in [140], uses the best two objective function values found so far. Let $Y_{(0)}^k = \min_{j=0,1,\ldots,k} f(X_j)$ be the smallest objective function value found after k iterations, and let $Y_{(1)}^k$ be the second smallest. The estimator of the global optimum after k iterations is given by,

$$\hat{f}(X_0, \ldots, X_k) = Y_{(0)}^k - \frac{Y_{(1)}^k - Y_{(0)}^k}{(1 - \alpha)^{-n/2} - 1}$$

which is the lower endpoint of a $100(1 - \alpha)\%$-confidence interval for y_*.

The approach taken by Romeijn and Smith [140, 139] is an important contribution towards developing an analytically derived cooling schedule. The main drawback is that the analysis depends on approximating the objective function with a quadratic function. The analytical approach has recently been extended to incorporate Lipschitz functions $f(x)$ by Kiatsupaibul and Smith [85, 86]. The full details are omitted here, but a sketch of the procedure is outlined. See Chapter 4 in [85] for the complete development.

Suppose now that the objective function $f(x)$ satisfies the Lipschitz condition with Lipschitz constant K. The goal is similar, given a current objective function value Y_k, and using a Boltzmann distribution with temperature T_k on S, find T_k such that,

$$P(f(X_{k+1}) < Y_k) \geq 1 - \alpha,$$

where $0 < \alpha < 1$. The development of the cooling schedule requires three basic steps.

The first step is to create a worst case function, denoted $h_0(x)$, such that the probability of sampling an improving point on the worst case function (i.e., using a Boltzmann distribution with function $h_0(x)$ and temperature T) is no greater than the probability of sampling an improving point on the original function (i.e., using a Boltzmann distribution with function $f(x)$ and temperature T)

$$P(f(X_{k+1}^f) < Y_k^f) \geq P(h_0(X_{k+1}^{h_0}) < Y_k^{h_0}),$$

where the superscript indicates the function used in the Boltzmann distribution.

The second step is to find a temperature T that maintains at least $1 - \alpha$ probability of sampling an improving point on the worst case function, that is, satisfies the inequality

$$P(h_0(X_{k+1}^{h_0}) < Y_k^{h_0}) \geq 1 - \alpha,$$

where $0 < \alpha < 1$. Then the temperature T is appropriate for a cooling schedule for adaptive search. However, when the number of iterations becomes large, the worst case function h_0 becomes complicated and it is difficult to solve for the temperature T.

The third step is to construct a simpler function $\hat{h}(x)$, called the hat function, which has a lower probability of generating an improving point, and can be used to calculate a cooling schedule.

To define the function $\hat{h}(x)$, we need to extend the notation used so far. As before, we have x_* is a global minimum, with $y_* = f(x_*)$. We assume the objective function $f(x)$ satisfies the Lipschitz condition with Lipschitz constant K. We sample X_{k+1} using a Boltzmann distribution on S with temperature parameter T, and because we are interested in defining a hat function, we now denote a Boltzmann distribution with two parameters, the function and the temperature,

$$\pi_{f,T}(x) = \frac{e^{-f(x)/T}}{\int_{x \in S} e^{-f(x)/T} dx}.$$

We also are interested in the probability of sampling a point in an improving region which depends on the function, so we now add a subscript to the notation for an improving set,

$$S_f(y) = \{x \in S : f(x) < y\}.$$

The hat function $\hat{h}(x)$ is constructed using the Lipschitz constant K, the global minimum x_* and $y_* = f(x_*)$, and the best value sampled thus

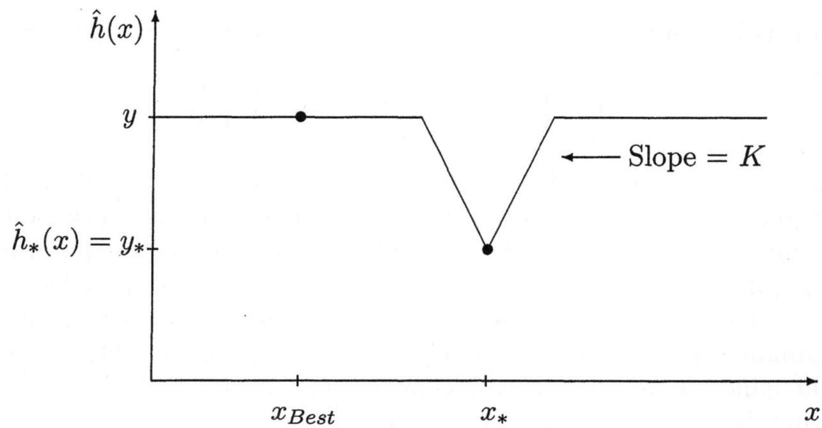

Figure 4.2. Illustration of the hat function $\hat{h}(x)$ in one dimension with Lipschitz constant K.

far, y. Let

$$\hat{h}(x) = \min\{y, g(x)\}$$

where

$$g(x) = y_* + K||x - x_*||$$

for all $x \in S$. The hat function is illustrated in Figure 4.2 for a one-dimensional problem. The hat function has the property that the probability of generating an improving point, with a fixed temperature T, for the original function is greater than that for the hat function,

$$\pi_{f,T}(S_f(y)) \geq \pi_{\hat{h},T}(S_{\hat{h}}(y)),$$

see [85, Corollary 4.1].

Now find T such that

$$\pi_{\hat{h},T}(S_{\hat{h}}(Y_k)) = 1 - \alpha.$$

There are two assumptions. One is that the level set at Y_k is in S. The other is that the Lebesgue measure of S (i.e., n-dimensional volume of S) is known. Upper and lower bounds on the temperature at the kth iteration are given as

$$\frac{(1-\epsilon)(f(Y_k) - y_*)}{L} \leq T_k \leq \left(\frac{M}{dM-1}\right)(f(Y_k) - y_*) \qquad (4.4)$$

where

$$L = \ln\left(\left(\frac{(1-\alpha)}{\alpha}\right)\left(\frac{\nu(S)-\nu(B_{\epsilon\delta})}{\nu(B_{\epsilon\delta})}\right)\right)$$

and ν is Lebesgue measure, $B_{\epsilon\delta}$ is an n-dimensional ball of radius $\epsilon \cdot \delta$, and

$$M = \left(\frac{K}{f(Y_k)-y_*}\right)^n \left(\frac{(1-\alpha)}{\alpha}\right)\frac{V}{C}$$

with $V = \nu\left(S \setminus S_{\hat{h}}\right)$ and $C = 2\pi^{n/2}/\Gamma(n/2)$ (see [85, Proposition 4.4]).

Equation 4.4 provides a temperature T_k that guarantees the cooling schedule for adaptive search. The information needed to compute the temperature are the dimension n, the objective function at the last point $f(Y_k)$, the Lipschitz constant K, the Lebesgue measure of S, and the minimum objective function value y_*. The dimension and $f(Y_k)$ are readily at hand. The other three pieces of information must be estimated, or bounds provided. This is an active research area, and perhaps other functions besides the hat function could be used to construct the cooling schedule. Nevertheless, this approach of analytically obtaining a cooling schedule for adaptive search provides a theory in place of guesswork.

4. Summary

Annealing adaptive search and adaptive search were defined in this chapter, with the assumption that a sample point can be drawn exactly from a Boltzmann distribution with parameter T. The number of record values of AAS (i.e., adaptive search iterates) are bounded by PAS, and hence inherit the linearity property of PAS. The number of iterations of AAS is bounded by a special case of HAS, and an expression for the upper bound is given. This gives rise to potential cooling schedules. The performance analysis for AAS also points to the need to develop methods that can sample approximately from a Boltzmann distribution.

Chapter 5

BACKTRACKING ADAPTIVE SEARCH

Backtracking adaptive search is a generalization of hesitant adaptive search that not only allows hesitation, but allows the algorithm to backtrack by accepting points with worse objective function values. Backtracking adaptive search is similar to AAS because it allows sampling from the full set S. It is a generalization of PAS and HAS, which both restrict the sampled points to be in the improving region. Backtracking adaptive search differs from AAS in the specific sampling distribution used. The sampling distribution of AAS is a Boltzmann distribution with a temperature parameter that depends on the best objective function value sampled thus far. The sampling distribution of backtracking adaptive search depends on the *current* objective function value, instead of the *best* objective function value. It modifies the underlying sampling distribution based on this current value. Backtracking adaptive search is a step toward more realistic algorithms, like simulated annealing, because it allows the sampling distribution to change as the algorithm accepts improving and non-improving points.

Two slightly different versions of backtracking adaptive search (BAS) have been introduced in the literature and are summarized in this chapter. The detailed proofs of results in this chapter can be found in the individual papers. Section 1 in this chapter describes the mixed BAS algorithm introduced by Wood and colleagues in [27, 179] which allows a general range distribution and can include both continuous and discrete variables in the global optimization problem. An expression for the expected number of mixed BAS iterations to achieve a specified level of accuracy was developed by Bulger, et al. [27]. The complete distribution for the number of mixed BAS iterations to convergence was developed by Wood, et al. [179] and is summarized in Section 1.

Section 2 describes a second version of BAS which is called discrete backtracking adaptive search and was introduced by Kristinsdottir, et al. [96]. The two versions of BAS are each more general than the other in some respects. The version presented in Section 1 allows a general range distribution capturing global optimization problems with mixed continuous and discrete variables, while the version presented in Section 2 is restricted to discrete problems. The first version is referred to as mixed BAS in this book, and the second version is called discrete BAS. The mixed BAS algorithm is more restrictive in that it requires the sampling distribution to be the normalized restriction of a single underlying distribution, while the discrete BAS algorithm allows a general sampling distribution.

While mixed BAS, like the algorithms already discussed (PAS, AAS, and HAS), assumes that the sampling distribution depends only on the objective function value and not on the specific point in the domain, discrete BAS differs in this regard. For discrete BAS, we use a Markov chain analysis where the states are considered points in the domain. This gives rise to potentially a very large matrix. However, under certain conditions, a range analysis captures the behavior of the Markov chain on the domain. The conditions under which the domain states can be lumped together are discussed in Section 2.1, and an example of discrete BAS is given to illustrate the concept. In Section 2.2, an embedded Markov chain approach is used to develop bounds on performance. Two examples illustrate the analysis and explore the impact of backtracking by varying the probability of accepting non-improving points.

1. Mixed Backtracking Adaptive Search (Mixed BAS)

The global optimization problem (P) in this section is taken to be very general, where S is assumed to be a measurable space, and $f : S \rightarrow \mathbb{R}$ is a measurable function. As in Chapter 3, the number of iterations to first sample an objective function value of y or less $N(y)$ is the primary measure of performance, but it is easier to work with $M(y)$ the number of iterations *just before* landing in the specified region.

The mixed backtracking adaptive search algorithm is a generalization of hesitant adaptive search, and is similarly governed by the underlying sampling distribution δ on the domain S and three probability functions; the worsening probability $w(y)$, the hesitating probability $h(y)$, and the bettering probability $b(y)$. The first sample point X_0 is selected according to δ on S. At each subsequent iteration, the algorithm may worsen, hesitate at the current objective function value y, or better the objective function value. It is assumed that the respective probabilities sum to

1, $w + h + b \equiv 1$. If the algorithm worsens, it then generates a point according to the normalized restriction of δ on the set of points with larger objective function values; if the algorithm betters, it generates a point according to the normalized restriction of δ on the set of improving points with smaller objective function values. Notice that HAS is a special case of mixed BAS occurring when $w = 0$, and that PAS is the special case when additionally $b = 1$.

In Chapter 3, an analogy was made between HAS and leaping down a ladder or sliding down a pole. The analogy with BAS can be extended where now the person may ascend, pause, or descend the ladder according to the probabilities $w(y), h(y)$, and $b(y)$. The distance traveled up or down depends on the sampling distribution δ restricted to the appropriate set. If a person climbs the ladder (i.e., backtracks), then the next movement depends on the adjusted probabilities for the current rung. This differs from AAS where a person may backtrack, but the sampling distribution depends on the best level achieved rather than the current level.

The mixed backtracking adaptive search algorithm, from [179], is stated more formally below.

Mixed Backtracking Adaptive Search (Mixed BAS)

Step 0. Initialize $X_0 \in S$ according to probability measure δ on S. Set $k = 0$. Set $Y_0 = f(X_0)$.

Step 1. Generate X_{k+1} according to the normalized restriction of δ on:

$$\begin{array}{ll} \{x \in S : f(x) > Y_k\} & \text{with probability of worsening } w(Y_k), \\ \{x \in S : f(x) = Y_k\} & \text{with probability of hesitating } h(Y_k), \\ \{x \in S : f(x) < Y_k\} & \text{with probability of bettering } b(Y_k). \end{array}$$

Set $Y_{k+1} = f(X_{k+1})$.

Step 2. If a stopping criterion is met, stop. Otherwise, increment k and return to Step 1.

The mixed BAS algorithm can be expressed as a Markov chain, which is fundamental to the analysis. First, recall the probability measure δ on S dictates the sampling distribution in the domain, and ρ is the range probability measure defined by $\rho(A) = \delta(f^{-1}[A])$ for each measurable subset A of $\mathbb{R}$. Also, p is the range cumulative distribution function associated with ρ, so for $y \in \mathbb{R}$,

$$p(y) = \rho((-\infty, y]) = \delta(S(y)),$$

where $S(y) = \{x \in S : f(x) \leq y\}$. This is consistent with the notation elsewhere in the book.

The one-step transition kernel for mixed BAS is given in [27] both in terms of the domain with δ and the range with ρ as follows,

$$\begin{aligned}\alpha(x,\cdot) &= w(f(x))\delta|_{f^{-1}[(f(x),\infty)]}\\ &+ h(f(x))\delta|_{f^{-1}[\{f(x)\}]}\\ &+ b(f(x))\delta|_{f^{-1}[(-\infty,f(x))]}\\ \beta(y,\cdot) &= w(y)\rho|_{(y,\infty)}\\ &+ h(y)\rho|_{[y]}\\ &+ b(y)\rho|_{(-\infty,y)}\end{aligned}$$

where $\delta|_C$ and $\rho|_C$ denote the normalized restriction of δ and ρ, respectively, to C. Intuitively, the probability of making a transition from a current point x or objective function level y depends on the probability of worsening, hesitating, and bettering, and also on the underlying sampling distribution normalized to the region associated with the current value.

The first observation, made in [179, Proposition 3.2], is that the record values of mixed BAS follow the same distribution as the objective function values of PAS. Intuitively this makes sense because the record values of mixed BAS are strictly improving as in PAS, and the backtracking affects the time between record values but does not affect the amount of improvement given a record occurs. This is similar to the result in Lemma 3.1 that the record values of HAS are equal in distribution to PAS, and is stated in the following lemma.

LEMMA 5.1 *The stochastic process $\{Y_{R(k)}, k = 0, 1, \ldots\}$ of mixed BAS records is equal in distribution to the stochastic process $\{W_k, k = 0, 1, \ldots\}$ of PAS objective function values, i.e.,*

$$\{W_k\} \sim \left\{Y_{R(k)}\right\}$$

when using the same range probability measure ρ.

Proof: The proof in [179] follows the same pattern used in Lemma 2.1 and Lemma 3.1. ∎

The primary result for mixed BAS [179, Theorem 3.3] is stated below in Theorem 5.2. It provides the distribution for the number of iterations to convergence. The development is similar to HAS in that the distribution is given in terms of the factorial moment-generating function.

The main difference between mixed BAS and HAS is the difference in the distribution of the time between records. In HAS, the time between records is roughly a geometric distribution parameterized by the bettering probability, while the time between records for mixed BAS includes the time spent backtracking. Wood, et al. [179, Proposition 3.1] provide the delay distribution for mixed BAS. The primary result stated in Theorem 5.2 includes the expression for the delay distribution as well as the distribution for the number of mixed BAS iterations to convergence.

The notation is consistent with Chapter 3, where the region prior to termination (y, ∞) is partitioned by a collection of non-empty open sets where p is continuous, and the points of discontinuity are p_i. There are at most countably many points of discontinuity of p, denoted by $\{y_i : i \in I^D\}$. Also let $\pi_i = \rho(\{y_i\})$, $p_i = p(y_i)$, $b_i = b(y_i)$ and $w_i = w(y_i)$ for each $i \in I^D$. The complement of $\{y_i : i \in I^D\}$ in (y, ∞) is a countable union of open intervals, and we index these intervals by I^C, so that $(y, \infty) = \bigcup_{i \in I^C} U_i \cup \bigcup_{i \in I^D} \{y_i\}$. If the number of discontinuities is finite, with the largest being y_K, then the number of open sets U_i is just the number of discontinuities plus one. The "top" one, say U_K,is (y_K, ∞). In general, we let $U_i = (a_i, b_i)$. Let $M(U_i)$ denote the number of iterations BAS spends with record in U_i and $M_i = M(\{y_i\})$ denote the number of iterations BAS spends with record value y_i.

THEOREM 5.2 *Let p be a general range distribution function and $\{y_i : i \in I^D\}$ and $\{U_i = (a_i, b_i) : i \in I^C\}$ be the associated partition of (y, ∞). Then for BAS with worsening function w and bettering function b, the factorial moment-generating function for $M(y)$ is given by*

$$\begin{aligned} M(y, z) &= E\left[z^{M(y)}\right] \\ &= \prod_{i \in I^C} \exp\left\{\int_{U_i} \frac{D(t, z) - 1}{p(t)} d\rho(t)\right\} \prod_{i \in I^D} \left\{1 + \frac{\pi_i}{p_i}(D(y_i, z) - 1)\right\} \end{aligned}$$

where the factorial moment-generating function of the delay at level t $D(t, z) = E\left[z^{M(t^-)} | Y_0 = t\right]$ is equal to

$$\frac{z(b(t)p(t)(1 - p(t)) + w(t)(p(t) - \rho(t))(M(t, z) - p(t)))}{p(t)(1 - p(t))(1 - z(1 - b(t) - w(t))) - z\rho(t)w(t)(M(t, z) - p(t))}.$$

Proof: See [179]. ∎

The following corollary provides an integral equation for $E[M(y)]$ the expected number of mixed BAS iterations before termination at level y.

This first appeared in [27, Theorem 1]. Loosely speaking, the theorem states that

$$E[M(y)]' = \frac{1 - p(y) + w(y)E[M(y)]}{(1 - p(y))\,(p(y)(b(y) + w(y)) - w(y)\rho(\{y\}))} p'(y) \qquad (5.1)$$

where the prime indicates differentiation with respect to y.

COROLLARY 5.3 *For every* $y \in \mathbb{R}$,

$$E[M(y)] = \int_{(y,\infty)} \frac{1 - p(t) + w(t)E[M(t)]}{(1 - p(t))\,(p(t)(b(t) + w(t)) - w(t)\rho(\{t\}))} d\rho(t).$$

Proof: See [27]. ∎

Several examples illustrate Corollary 5.3. First, if $w(y) = 0$ for all y, then mixed BAS reduces to HAS because there is no backtracking, as noted in [27, Example 2]. Setting $w(y) = 0$ in Corollary 5.3 yields the expression for the expected number of HAS iterations to convergence given in Corollary 3.7,

$$E[M(y)] = \int_{(y,\infty)} \frac{d\rho(t)}{b(t)p(t)}.$$

As another example, mixed BAS reduces to pure random search when $w(y) = 1 - p(y)$ and $b(y) = p(y) - \rho(\{y\})$ for all $y \in \mathbb{R}$ [27, Example 3]. In this case, the expression in Corollary 5.3 becomes

$$E[M(y)] = \int_{(y,\infty)} \frac{1 + E[M(t)]}{p(t) - \rho(\{t\})} d\rho(t).$$

This equation has the solution $E[M(y)] = 1/p(y) - 1$, which agrees with the result given for PRS in Chapter 2, Equation 2.2. The difference in the extra 1 can be explained by the difference between $N(y)$, the number of iterations to get below level y, and $M(y)$, the number of iterations *before* getting below level y.

Another example sets ρ to be the uniform distribution on $(0, 1)$ so that $p(y) = y$ on $(0, 1)$ [27]. Also, suppose BAS always has a 50% chance of improving or backtracking, that is, $b(y) = w(y) = 1/2$ for all $y \in (0, 1)$. In this example ρ is continuous, and $\rho(\{t\}) = 0$ for any singleton $\{t\}$, so Equation 5.1 has the general solution

$$E[M(y)] = \sqrt{\frac{1 - y}{y}} \left(C + \sin^{-1}(1 - 2y)\right).$$

In the limit, as $y \to 1$ with $E[M(y)] \to 0, p(y) \to 1$ and $p'(y) \to 1$, so we must have $C = \pi/2$, which yields the expected number of mixed BAS iterations before a value not greater than y is sampled as

$$E[M(y)] = \sqrt{\frac{1-y}{y}} \left(\pi/2 + \sin^{-1}(1-2y)\right).$$

In this section, we have summarized a complete analysis of mixed BAS. Mixed BAS is defined by its underlying sampling distribution as well as its worsening, hesitating, and bettering probabilities. The global optimization problem can involve both continuous and discrete variables, as ρ may be a general distribution. The analysis completely characterizes the progress of the objective function values $\{Y_k\}$ of mixed BAS through the distribution of $M(y)$.

2. Discrete Backtracking Adaptive Search (Discrete BAS)

We next investigate discrete BAS which assumes that the global optimization problem (P) has a finite set of distinct objective function values. Discrete BAS, as defined by Kristinsdottir, et al. [96], is characterized by an underlying sampling distribution called the generating probability and an acceptance probability. It differs from mixed BAS in that the sampling probabilities are not defined by normalizing an underlying distribution on the improving or worsening regions.

Discrete BAS involves three factors that affect its rate of convergence: the generator, the acceptance probability and the landscape of the function being optimized. Convergence properties of simulated annealing were analyzed in [1] for combinatorial optimization problems using a Markov chain approach. In [1], asymptotic results of convergence were given, whereas here we are interested in the number of iterations until first sampling the global optimum. This section follows a Markov chain approach. Markov chains have been used to study simulated annealing [1] as well as to study the convergence properties genetic algorithms [144, 40].

To define the discrete BAS algorithm, let $x_1, x_2, \ldots, x_M$ denote the M points in the domain S. Also let $y_i = f(x_i)$ for $i = 1, \ldots, M$. The generator is described by a probability transition matrix G where g_{ij} is the probability that point x_j is generated from the current point x_i. Similarly the acceptance criterion is described by an acceptance probability matrix A, where a_{ij} is the probability that the candidate point x_j is accepted given the current point x_i. Consequently, the probability that discrete BAS moves from point x_i to x_j in one iteration is $p_{ij} = g_{ij}a_{ij}$.

The discrete backtracking adaptive search algorithm, from [96], is stated as follows.

Discrete Backtracking Adaptive Search (Discrete BAS)

Step 0. Initialize $X_0 \in S$ according to probability measure δ on S. Set $k = 0$. Set $Y_0 = f(X_0)$.

Step 1. Generate a candidate point V_{k+1} according to probability matrix G.

Step 2. Update X_{k+1} according to the acceptance probability matrix A where

$$X_{k+1} = \begin{cases} V_{k+1} & \text{with acceptance probability } a_{ij} \\ X_k & \text{otherwise} \end{cases}$$

and a_{ij} is determined by $x_i = X_k$ and $x_j = V_{k+1}$. Set

$$Y_{k+1} = \begin{cases} f(X_{k+1}) & \text{if } f(X_{k+1}) < Y_k \\ Y_k & \text{otherwise} \end{cases}$$

and set $X_{\text{best}} = X_{k+1}$ if $f(X_{k+1}) < Y_k$.

Step 3. If a stopping criterion is met, stop. Otherwise, increment k and return to Step 1.

Discrete BAS is able to capture many types of generators, including neighborhood specific generators, using a probability matrix G to describe the sampling distribution. The generating probabilities g_{ij} may be zero for points x_j far from x_i, but to be well-defined g_{ij} must sum to one across rows, i.e., $\sum_{j=1}^{M} g_{ij} = 1$ for all $i = 1, \ldots, M$.

The acceptance probability matrix A, used in Step 2, may include a variety of distributions, but we assume that the candidate point V_{k+1} is always accepted if it is improving or of equal objective function value to the current point, that is, $a_{ij} = 1$ if $f(x_j) \leq f(x_i)$. The acceptance probability can reflect the difference in objective function values between the current point and the candidate point. For instance, define a_{ij} as

$$a_{ij} = \exp\left(\frac{f(x_i) - f(x_j)}{T}\right)$$

for $f(x_j) > f(x_i)$. This is the Metropolis criterion common in simulated annealing (see Equation 1.5), where a large difference in objective function values yields a small acceptance probability. The main difference between discrete BAS and simulated annealing is that for discrete BAS

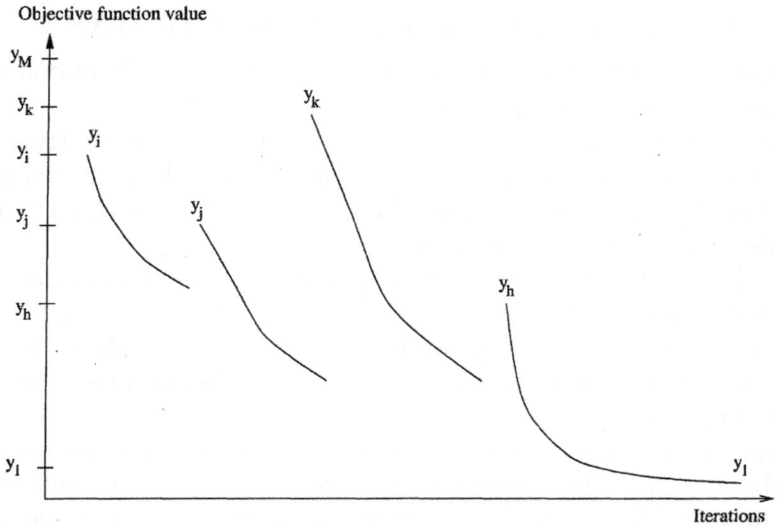

Figure 5.1. Series of improving points showing acceptance of non-improving points. A curve consists of a downward run, together with the first higher value.

the probability of accepting a non-improving point is constant for all iterations of the algorithm, whereas in simulated annealing the acceptance probability changes according to a cooling schedule. For example, a cooling schedule of $T_k = 0.9^k$ may be used in simulated annealing, but is not included in discrete BAS because it involves the iteration number. If simulated annealing uses a temperature parameter T_k that depends only on the objective function value and not the iteration value, then discrete BAS can be used to describe that algorithm.

The analysis of the stochastic adaptive search algorithms in this book has emphasized the progress of improving points, or record values observed during the algorithm. The analogy of leaping down a ladder or sliding down a pole was used earlier to describe the backtracking, hesitating or improving behavior of an algorithm. Figure 5.1 provides another illustration of how backtracking adaptive search or simulated annealing progresses. The algorithm generates a series of points, and sometimes the points may be improving, and sometimes the algorithm may accept a non-improving point. The analysis of discrete BAS groups the sequence of improving points ending with acceptance of a non-improving point as a "curve." In Figure 5.1 there are four curves. Given a starting point, each curve is independent of each other, which aids in the analysis. This is discussed in Section 2.2, where the analysis of discrete BAS uses embedded Markov chains.

2.1 Markov Chain Models of Discrete BAS

To model discrete BAS and analyze the number of iterations until convergence, we define a Markov chain. We can set up two different Markov chain models, one on the domain S, and the other on the range of objective function values. The domain Markov chain allows more detail than the range Markov chain, but the range Markov chain is more tractable to derive bounds on performance.

In setting up the Markov chain model on the domain, consider the M distinct points in S, $x_1, x_2, \ldots, x_M$, as the M states of a Markov chain. We model the global optimum x_1 as an absorbing state, with all other states being transient. It is possible to include multiple optima by creating more absorbing states.

The one-step transition matrix P describes the algorithm, where p_{ij} is the probability that the algorithm moves from x_i to x_j in a single iteration. The transition probabilities for the absorbing state corresponding to the global optimum x_1 are simply $p_{11} = 1$ and $p_{1j} = 0$ for $j \neq 1$. For the rest of the states, $i = 2, \ldots, M$, the transition probabilities are determined by the generating and accepting probabilities. Given that the algorithm is at a point x_i, the probability that point x_j is generated is g_{ij}, and the candidate point is accepted with probability a_{ij}, hence the transition probability is $p_{ij} = g_{ij}a_{ij}$ for $i \neq j$. The algorithm stays at the same point x_i if either it is generated again, or if another point is generated and rejected, therefore $p_{ii} = g_{ii} + \sum_{j \neq i} g_{ij}(1-a_{ij})$.

For the analysis, it is notationally convenient to assume that the M points in the domain have unique objective function values with $f(x_1) < f(x_2) < \ldots < f(x_M)$. If we also assume that improving points are always accepted, i.e., $a_{ij} = 1$ for $j < i$, then the one-step transition matrix has a triangular pattern. The elements p_{ij} for $j < i$ indicate transitions to improving points and $p_{ij} = g_{ij}$. The elements p_{ij} for $j > i$ indicate backtracking, or accepting non-improving points, and $p_{ij} = g_{ij}a_{ij}$. The diagonal elements p_{ii} may be interpreted as hesitating, either by generating the same point again or by rejecting a candidate point, therefore $p_{ii} = g_{ii} + \sum_{j=i+1}^{M} g_{ij}(1-a_{ij})$. Figure 5.2 shows the transition matrix P for the domain Markov chain in this case.

The expected number of iterations to first sample the global optimum $E[N(1)]$ is equal to the expected number of iterations to absorption, and can be expressed in terms of the transition matrix of the Markov chain [82]. In this section, we modify the notation slightly. Instead of using $E[N(y)]$ to specify an objective function level y, we let $E[N_i]$ denote the expected number of iterations to absorption in state x_1, representing the global optimum, given that the initial starting state was x_i, where $i = 2, \ldots, M$. The expected number of iterations to absorption indicates

$$
\begin{array}{c}
\\ M \\ \vdots \\ i \\ \vdots \\ 1
\end{array}
\begin{array}{c}
\begin{array}{ccccc} M & & i & & 1 \end{array} \\
\left[\begin{array}{ccccc}
g_{MM} & \cdots & g_{Mi} & \cdots & g_{M1} \\
\vdots & \vdots & \vdots & \vdots & \vdots \\
g_{iM}a_{iM} & \cdots & g_{ii}+\sum_{j=i+1}^{M} g_{ij}(1-a_{ij}) & \cdots & g_{i1} \\
\vdots & \vdots & \vdots & \vdots & \vdots \\
0 & \cdots & 0 & \cdots & 1
\end{array}\right]
\end{array}
$$

Figure 5.2. Entries in the one-step transition matrix for a domain Markov chain with ordered states.

the average computational effort to first sample the global optimum, but not necessarily to confirm it. If the algorithm starts at the worst point x_M, then $E[N_M]$ is comparable to $E[N(1)]$ used elsewhere in this book. The following theorem from [82], states how the expected number of iterations to absorption depends on P.

THEOREM 5.4 *The expected number of iterations to absorption in state x_1, starting in state x_i, $i = 2, \ldots, M$, can be found by solving the following system of equations,*

$$E[N] = (I - Q)^{-1}e \tag{5.2}$$

where Q is an $(M-1) \times (M-1)$ matrix that is the transient part of the one-step transition matrix P, I is the $(M-1) \times (M-1)$ identity matrix and e is a $M-1$ vector of ones.

Proof: The proof is in [82]. ∎

Theorem 5.4 provides a method to calculate the expected number of iterations to absorption (conditional on a given starting state) using the matrix $(I - Q)^{-1}$, often referred to as the fundamental matrix. The variance of the number of iterations to absorption is also obtainable from the fundamental matrix $(I - Q)^{-1}$ (see [82, Theorem 3.3.3]).

The domain Markov chain formulation is very general and can be used to model a variety of random search algorithms on various problem domains. However the $I - Q$ matrix may be extremely large, with no apparent structure to aid the inversion. This leads to the development of the range Markov chain model, and later to the embedded range Markov chain model.

The range Markov chain model simplifies the analysis, at the cost of restricting the type of generator and function that can be analyzed. The main difference between the range Markov chain and the domain Markov chain is the interpretation of states. Intuitively, one could lump all the points in the domain with equal objective function values into one state. Therefore, in the range Markov chain, state i represents the set of points with objective function value y_i. If each point in the domain has a unique objective function value, then there are M states in both the domain Markov chain and the range Markov chain, and in this case, the one-step transition matrices are identical (as shown in Figure 5.2). If there are multiple points with equal objective function value, then the range Markov chain has fewer states. The issue remains of how to define the one-step transition matrix while maintaining an accurate description for the behavior of the algorithm. Kemeny and Snell [82] compare an original Markov chain with a reduced state Markov chain called the *lumped Markov chain*, and provide necessary and sufficient conditions for when the lumped Markov chain is equivalent to the original Markov chain with respect to coarser properties, including the expected number of iterations to absorption. The expected number of iterations to absorption for the range Markov chain model can be found by using the result in Theorem 5.4.

We reduce the number of states by *lumping* domain states by their range values. To distinguish between the domain Markov chain and the range Markov chain, we use the notation $x_1, \ldots, x_M$ for the states in the domain Markov chain, and $y_1, \ldots, y_M$ for the states in the range Markov chain. We create a partition of the domain Markov chain by determining the set of inverse images $f^{-1}(y_j)$ of states for the range Markov chain. Then state y_j in the range Markov chain denotes the partition set of points with objective function value y_j, and the range Markov chain has states $y_1, \ldots, y_M$. The number of states in the range Markov chain is at most M, corresponding to the case where all points have unique objective function values. Typically we expect the number of states in the range Markov chain to be less than M. The probability of making a transition from an original domain state x_i to a partition set y_j can be expressed as $\tilde{p}_{x_i y_j} = \sum_{x_k \in y_j} p_{x_i x_k}$, where a domain state x_k is in partition set y_j if $f(x_k) = y_j$.

As defined in [82, page 124], a Markov chain is lumpable with respect to a partition if for every starting vector the lumped process is also a Markov chain and the transition probabilities do not depend on the starting vector. In addition [82, Theorem 6.3.2], a necessary and sufficient condition for a Markov chain to be lumpable with respect to a partition is that for every pair of sets in the partition, A and B, the transition probability from any original state in set A to the set B has the same value for every original state in set A. In the context of partitioning the domain Markov chain by objective function value, a necessary and sufficient condition for lumpability is that, for every pair of partition sets y_i and y_j, the probability $\tilde{p}_{x_l y_j}$ must have the same value for every state x_l corresponding to y_i. We denote these common values $\hat{p}_{ij} = \tilde{p}_{x_l y_j}$ for any $x_l \in y_i$. These values define the transition probabilities for the *lumped* range Markov chain. Thus, if the domain Markov chain process is lumpable with respect to the partition set of distinct objective function values, then the domain Markov chain model can be reduced (lumped) to give an equivalent range Markov chain model.

Example 1. We illustrate the idea of lumpability with an example from [96]. First consider discrete BAS on a problem that is modeled by the domain Markov chain model. In this example, there are seven points in the domain, two with objective function value four, two with objective function value three, two with objective function value two and one global optimum with objective function value one. The one-step transition matrix for this problem with the states representing the seven points in the domain is shown in the top portion of Figure 5.3. Parameter t is the probability of accepting any non-improving point. The probability of accepting a non-improving point may be interpreted as corresponding to a fixed temperature parameter in a simulated annealing algorithm. Later in Section 2.3 the impact of varying t is explored.

Now we verify that this Markov chain is lumpable on the problem domain, giving the range Markov chain shown in the bottom part of Figure 5.3. Notice that the necessary and sufficient conditions for lumpability for any pair of partition sets y_i and y_j are satisfied. For instance, $\tilde{p}_{x_{4*} y_4} = \sum_{x_k \in y_4} p_{x_{4*} x_k} = p_{x_{4*} x_4} + p_{x_{4*} x_{4*}} = 0.07 + 0.03 = 0.1$, while $\tilde{p}_{x_4 y_4} = \sum_{x_k \in y_4} p_{x_4 x_k} = p_{x_4 x_4} + p_{x_4 x_{4*}} = 0.03 + 0.07 = 0.1$. The conditions of lumpability are satisfied for the other states in the domain Markov chain as well. These two models are equivalent in terms of the distribution of the number of iterations to absorption, and hence the domain and range processes have the same expected number of iterations to first find the global optimum.

2.2 Range embedded Markov chain model

Using the range Markov chain model and Theorem 5.4, it is possible to obtain the expected number of iterations to absorption. However, without additional structure, the range $I - Q$ matrix may be full and

$$\begin{array}{ccccccc} 4 & 4^* & 3 & 3^* & 2 & 2^* & 1 \end{array}$$
$$\begin{bmatrix} 0.03 & 0.07 & 0.05 & 0.15 & 0.1 & 0.2 & 0.4 \\ 0.07 & 0.03 & 0.15 & 0.05 & 0.2 & 0.1 & 0.4 \\ 0.1t & 0.15t & 0.35-0.25t & 0.15 & 0.1 & 0.15 & 0.25 \\ 0.15t & 0.1t & 0.15 & 0.35-0.25t & 0.15 & 0.1 & 0.25 \\ 0.1t & 0.3t & 0.1t & 0.2t & 0.75-0.7t & 0.15 & 0.1 \\ 0.3t & 0.1t & 0.2t & 0.1t & 0.15 & 0.75-0.7t & 0.1 \\ 0 & 0 & 0 & 0 & 0 & 0 & 1 \end{bmatrix}$$

Domain Markov chain one-step transition matrix

$$\begin{array}{cccc} 4 & 3 & 2 & 1 \end{array}$$
$$\begin{bmatrix} 0.1 & 0.2 & 0.3 & 0.4 \\ 0.25t & 0.25+(1-t)0.25 & 0.25 & 0.25 \\ 0.4t & 0.3t & 0.2+(1-t)0.7 & 0.1 \\ 0 & 0 & 0 & 1 \end{bmatrix}$$

Range Markov chain one-step transition matrix

Figure 5.3. Domain and range transition matrices for Example 1 demonstrating lumpability.

therefore difficult to invert analytically. The motivation of the embedded Markov chain model, developed in [96], is to convert the range transition matrix into a form that is easier to analyze. Notice that if only improving points are accepted by an algorithm, then the resulting range transition matrix is upper triangular so can in many instances be inverted analytically. Discrete BAS can be modelled as a series of embedded Markov chains, which involve transition matrices with a triangular pattern. This is used, following [96], to derive upper and lower bounds for the expected number of iterations to find the optimum.

Referring back to Figure 5.1, discrete BAS produces a series of improving points that are always accepted, and then occasionally accepts a non-improving point. Each series of improving points, shown as a curve in the figure, corresponds to an embedded Markov chain. The transition matrix for the embedded Markov chain is then upper triangular because it includes only improving points. The accepted non-improving point then terminates one curve and starts another.

To model the algorithm with a series of embedded Markov chains, the M states of the Markov chain represent the objective function values, $y_1, \ldots, y_M$. The global optimum y_1 is an absorbing state and will be denoted 1, while the transient states include $2, \ldots, M$. To model acceptance of non-improving points, added to the model are distinct absorbing states, $2', \ldots, M'$ that reflect the event of accepting a non-improving point. That is, when a non-improving point j is accepted the chain reaches an absorbing state j'. A curve is formed when the chain reaches an absorbing state.

To describe this associated one-step transition matrix of the embedded Markov chain, illustrated in Figure 5.4, suppose the process is in a transient state i, where $i \in \{2, \ldots, M\}$. Three types of points might be sampled. First, the global optimum could be sampled, in which case the Markov chain is absorbed into state 1. Because this is always an improving point, the transition probability is just the probability that the global optimum is sampled, $p_{i1} = g_{i1}$, for $i \in \{2, \ldots, M\}$. Second, an intermediate non-absorbing non-worsening point could be sampled, and in this case it is always accepted, so $p_{ij} = g_{ij}$ for $1 < j \leq i$ and $i \in \{2, \ldots, M\}$. Third, a non-improving point y_j could be sampled, with probability g_{ij}, for $j > i$. Then this point is either accepted, with probability a_{ij}, or rejected. If it is accepted, the Markov chain moves to absorbing state j', and the transition probability is $p_{ij'} = g_{ij}a_{ij}$. If the point is rejected, the Markov chain stays at its current point i. Thus the transition probability from state i to i, for $i \in \{2, \ldots, M\}$, is the chance it is sampled directly plus the sum of probabilities that higher points are sampled and rejected, or $p_{ii} = g_{ii} + \sum_{k=i+1}^{M} g_{ik}(1 - a_{ik})$.

Figure 5.4 illustrates the one-step transition matrix P for the embedded Markov chain. In the top portion of Figure 5.4, the structure of the one-step transition matrix is separated into blocks, where Q^* includes transitions to the same or improving points and R^* includes transitions to non-improving points. The middle portion of the figure details Q^* and the bottom portion of the figure details R^*. Notice that Q^* is an upper triangular matrix and R^* is a lower triangular matrix. Both are $(M-1) \times (M-1)$ matrices.

The elements of Q^* for $i \in \{2, \ldots, M\}$ and $j \in \{2, \ldots, M\}$ are

$$q^*_{ij} = \begin{cases} 0 & \text{if } i < j \\ g_{ii} + \sum_{k=i+1}^{M} g_{ik}(1 - a_{ik}) & \text{if } i = j \\ g_{ij} & \text{if } i > j \end{cases} \tag{5.3}$$

and the elements of R^* for $i \in \{2, \ldots, M\}$ and $j' \in \{2', \ldots, M'\}$ are

$$r^*_{ij'} = \begin{cases} g_{ij}a_{ij} & \text{if } j > i \\ 0 & \text{if } j \leq i. \end{cases} \tag{5.4}$$

The following theorem from [96, Theorem 2] provides expressions for the expected number of iterations to find the global optimum when the algorithm is modeled using the embedded Markov chain.

$$
\begin{array}{c}
 \\ M \\ \vdots \\ 2 \\ 1 \\ M' \\ \vdots \\ 2'
\end{array}
\begin{array}{c}
\begin{array}{ccc} M \quad\quad 2 & 1 & M' \quad\quad 2' \end{array} \\
\left[\begin{array}{c|c|c}
Q^* & g_{i1} & R^* \\ \hline
0 & 1 & 0 \\ \hline
0 & 0 & I
\end{array}\right]
\end{array}
$$

Structure of one-step transition matrix, P

$$
\begin{array}{c|cccccc}
 & M & & i & & & 2 \\
\hline
M & g_{MM} & \cdots & g_{Mi} & & \cdots & g_{M2} \\
\vdots & \vdots & \vdots & \vdots & & \vdots & \vdots \\
i & 0 & \cdots & g_{ii} + \sum_{j=i+1}^{M} g_{ij}(1-a_{ij}) & & \cdots & g_{i2} \\
\vdots & \vdots & \vdots & \vdots & & \vdots & \vdots \\
2 & 0 & \cdots & 0 & & \cdots & g_{22}
\end{array}
$$

Structure of Q^*

$$
\begin{array}{c|ccccc}
 & M' & & i+1' & & 2' \\
\hline
M & 0 & \cdots & 0 & \cdots & 0 \\
\vdots & \vdots & \vdots & \vdots & \vdots & \vdots \\
i & g_{iM}a_{iM} & \cdots & g_{ii+1}a_{ii+1} & \cdots & 0 \\
\vdots & \vdots & \vdots & \vdots & \vdots & \vdots \\
2 & g_{2M}a_{2M} & \cdots & g_{2i+1}a_{2i+1} & \cdots & 0
\end{array}
$$

Structure of R^*

Figure 5.4. Structure of one-step transition matrix for the embedded Markov chain.

THEOREM 5.5 *The expected number of iterations to absorption in state 1, starting in state i, $i = 2, \ldots, M$ is given by the following system of equations*

$$E[N] = (I - F^*)^{-1} u \tag{5.5}$$

where F^* and u can be found by solving the following system of equations,

$$u = (I - Q^*)^{-1} e \tag{5.6}$$

$$F^* = (I - Q^*)^{-1} R^* \tag{5.7}$$

where Q^* and R^* are defined in Equations 5.3 and 5.4.

Proof: The proof is given in [96]. ■

Theorem 5.5 presents an alternative set of equations from Theorem 5.4 which can be used to solve for the expected number of iterations to find the optimum. The advantage to solving the three systems of equations in Theorem 5.5 for the embedded Markov chain is that the matrices have a special structure. The matrix inversions in Equations 5.6 and 5.7 are relatively easy because Q^* is upper triangular, but the matrix $(I - F^*)$ in Equation 5.5 is a full matrix which makes an analytical inversion extremely difficult. However, it is possible to get an upper and a lower bound for the expected number of iterations $E[N]$ by using Equations 5.6 and 5.7 which only require the inversion of an upper triangular matrix.

The following theorem from [96, Theorem 3] gives an upper bound and a lower bound for the expected number of discrete BAS iterations to first find the global optimum.

THEOREM 5.6 *The expected number of iterations to first sample the global optimum, starting in state i is bounded above by*

$$u_i + \max_{i \in \{2,\ldots,M\}} \left(\frac{u_i}{f_{i1}} \right) (1 - f_{i1}) \tag{5.8}$$

and is bounded below by

$$u_i + \min_{i \in \{2,\ldots,M\}} \left(\frac{u_i}{f_{i1}} \right) (1 - f_{i1}) \tag{5.9}$$

where u_i and f_{i1} are as defined in equations (5.6) and (5.7).

Proof: The proof is given in [96]. ■

The upper and lower bounds are expressed in terms of u_i and f_{i1}. Here u_i is the expected number of iterations to absorption in any absorbing state, given that the algorithm started in state i, and f_{i1} is the probability of being absorbed in state 1 (the global optimum), given that the algorithm started in state i. Deriving u_i and f_{i1} only requires an inversion of an upper triangular matrix.

2.3 Examples of Discrete BAS

We now apply the results in Theorems 5.5 and 5.6 to two discrete BAS algorithms with specific generators and acceptance probabilities. This analysis was first presented in [96].

Example 1 in Section 2.1, which was used to illustrate a case where the conditions for lumpability are satisfied, is now used to illustrate the impact of accepting non-improving points on the performance of the algorithm. Example 1 depicts a situation where the probability of sampling the global optimum decreases the closer the objective function value is to the global optimum. This may reflect a ridge surrounding the global optimum. We analyze how a constant acceptance probability affects the expected number of iterations to first find the global optimum. In this example, a positive probability of accepting a non-improving point improves performance.

The second example in this section expands the combined PRS-PAS algorithm described in Chapter 3, Section 4. In Chapter 3, the combination of PRS-PAS was examined to see the effect that steady improvement had on the number of iterations to convergence. Now we add an acceptance probability to demonstrate the impact of accepting a worse point and losing the advantage PAS has on the expected number of iterations to convergence. We also derive upper and lower bounds for the expected number of iterations for both examples, and demonstrate that the exact expected number of iterations can be close to either bound.

Example 1 revisited. Recall the example of discrete BAS from Section 2.1. It had seven points in the domain, two with objective function value four, two with objective function value three, two with objective function value two and one global optimum with objective function value one. The one-step transition matrices for this problem for both the domain and range Markov chains are given in Figure 5.3. The parameter t is the probability of accepting any non-improving point. In this example, the probability of sampling the global optimum decreases the closer we are in objective function value to the global optimum. Notice that at objective function value 4 there is a fairly high probability of sampling the global optimum, 0.4, while at objective function value 3 this probability is lower, 0.25, and even lower at objective function value 2, where it is 0.1. This suggests that there may be an advantage to accepting non-improving points, because it might lead to a more likely path.

Theorems 5.5 and 5.6 are used to calculate the exact expected number of iterations to convergence, as well as upper and lower bounds for this example. The results are presented in Figure 5.5 for starting points 4, 3, or 2. The figure shows that including acceptance probability clearly

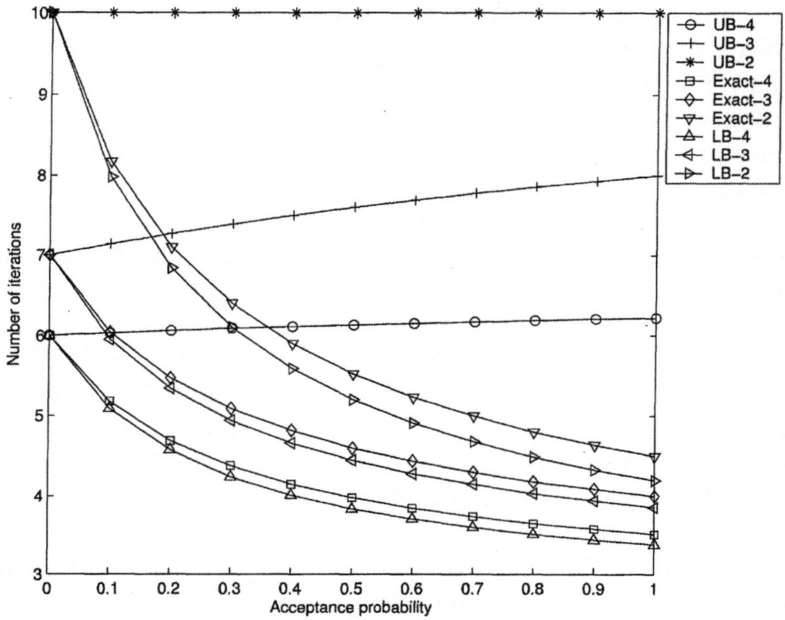

Figure 5.5. Upper bound (UB), lower bound (LB), and exact expected number of iterations to convergence for Example 1, for starting points of 4, 3 or 2.

helps the algorithm find the optimum in fewer iterations. For this type of generator and "landscape" it is beneficial to accept a non-improving point. The second fact to notice is how close the exact expected number of iterations is to the lower bound for the expected number of iterations.

Example 2: Combined PRS-PAS algorithm with acceptance probability. The combination of PRS and PAS presented in Chapter 3, Section 4 was used to demonstrate the power of steady improvement. The combined algorithm samples according to pure random search with probability $1-p$, and according to PAS with probability p, both based on the same generating distribution. Theoretically, it is most effective to have $p = 1$, but it may not be possible to implement. It was demonstrated in Chapter 3 and in [186] that only a small value of p can make a large improvement in performance. The analysis did not allow acceptance of non-improving points (i.e., $a_{ij} = 0$), and here we extend the previous analysis by using the embedded Markov chain to capture the effects of allowing a positive probability t of accepting a non-improving point. It is to be expected that allowing a positive probability of accepting a non-improving point will degrade performance. This provides a way to see the amount of increased computation needed when not able to always generate an improving point.

There are two versions of the PAS algorithm for discrete domains, a strong version and a weak version (see Chapter 2, Section 2). In this example we use the weak version of PAS where the improving set is $S(X_k) = \{x : x \in S \text{ and } f(x) \leq f(X_k)\}$. As before let $y_1 < y_2 < \ldots < y_M$ be distinct objective function values. In keeping with earlier notation, pure random search samples the domain according to a fixed probability distribution δ on S, independently of previous points. Given this sampling distribution, a probability measure $\pi = (\pi_1, \ldots, \pi_M)$ is defined on the range of f as follows. Let π_j be the probability that any iteration of pure random search attains a value of y_j, that is, $\pi_j = P(Y_0 = y_j) = \mu(f^{-1}(y_j))$ for $j = 1, \ldots, M$. Here p_j denotes $\sum_{i=1}^{j} \pi_i$, the probability that PRS attains a value of y_j or less. Given the current objective function level y_i, the probability PAS samples y_j, for $j < i$, is π_j / p_i.

The embedded Markov chain model is used to analyze this combined PRS-PAS algorithm that has non-zero probabilities a_{ij} of accepting non-improving points. The transition probability incorporates the probability of sampling according to PRS or PAS as well as the probability of accepting the sampled point even if it is not improving. This one-step transition matrix has the same structure as in Figure 5.4. The entries in the Q^* matrix, for $i = 2, \ldots, M$, and $j = 1, \ldots, M$ are

$$q^*_{ij} = \begin{cases} 0 & \text{if } i < j \\ \sum_{k=i+1}^{M} (1-p)\pi_k(1-a_{ik}) + (1-p)\pi_j + p(\frac{\pi_j}{p_i}) & \text{if } i = j \\ (1-p)\pi_j + p(\pi_j / p_i) & \text{if } i > j \end{cases} \tag{5.10}$$

and the entries in R^*, for $i = 2, \ldots, M$, and $j' = 2', \ldots, M'$ are

$$r^*_{ij'} = \begin{cases} (1-p)\pi_j a_{ij} & \text{if } i < j \\ 0 & \text{if } i \geq j. \end{cases} \tag{5.11}$$

When j' is strictly larger than i, the only way the algorithm can move from state i to j' is if state j' is sampled according to pure random search, and the point is accepted, thus $p_{ij'} = (1-p)\pi_j a_{ij}$. When $i = j$ the algorithm stays at i by sampling a non-improving point using PRS and rejecting it, or by sampling the current point again with PAS or PRS. When j is strictly less than i, then state j can be sampled according to PRS or PAS, thus $p_{ij} = (1-p)\pi_j + p(\pi_j / p_i)$, and the improving point is always accepted.

Next consider the special case where a uniform distribution is used for the generator and the acceptance probability t is constant. Assuming a uniform distribution gives $\pi_j = 1/M$ and $p_i = \sum_{k=1}^{i} \pi_k = i/M$, and a constant acceptance probability gives $a_{ij} = t$ for $j > i$. Using these assumptions, Equations 5.10 and 5.11 simplify, and the resulting transition matrix is shown in Figure 5.6.

Applying Theorem 5.6 to this combination of $(1-p)$PRS + pPAS algorithm with a uniform sampling distribution and acceptance probability t leads to the following theorem giving upper and lower bounds for the expected number of iterations $E[N_M]$, to first find the global optimum starting in state M. This was derived in [96, Theorem 4]. In particular, the number of iterations until convergence is stated in terms of p the probability of generating a point in the improving region using PAS and t the probability of accepting a non-improving point. The proof is not given here, but is available in [96]. Because M is the hardest initial state, finding the bounds on $E[N_M]$ does not require an inversion of $I - F^*$, it requires only an inversion of $I - Q^*$ to find u_M and f_{1M}. This can be done analytically because, for the embedded Markov chain model, $I - Q^*$ is upper triangular and has a special structure. The lower bound can be found similarly, using $E[N_2] \geq \frac{u_2}{f_{21}}$.

THEOREM 5.7 *The expected number of iterations to solve the global optimization problem for the combined PRS-PAS algorithm with probability p, assuming a uniform distribution for the generating method and a constant acceptance probability t, and starting in state M, is bounded above by*

$$\frac{1 + \sum_{j=2}^{M-1} \frac{j}{\rho_j} \prod_{k=j+1}^{M-1} (1 + \frac{\gamma_k}{\rho_k})}{\frac{\gamma_M}{\rho_M} \left(1 + \sum_{j=2}^{M-1} \frac{\gamma_j}{\rho_j} \prod_{k=j+1}^{M-1} (1 + \frac{\gamma_k}{\rho_k})\right)}$$

and is bounded below by

$$\frac{2M}{pM + 2(1-p)}$$

where γ_k and ρ_k are defined as follows,

$$\gamma_k = pM + k(1-p)$$

$$\rho_k = (k-1)\left(pM + k(1-p) + (\frac{k}{k-1})(1-p)(M-k)t\right).$$

Proof: See [96]. ■

$$
\begin{array}{c|c|c|c}
 & M \quad \cdots \quad 2 & 1 & M' \quad \cdots \quad 2' \\
\hline
\begin{array}{c} M \\ \vdots \\ 2 \end{array} & Q^* & \frac{1-p}{M} + \frac{p}{i} & R^* \\
\hline
1 & 0 & 1 & 0 \\
\hline
\begin{array}{c} M' \\ \vdots \\ 2' \end{array} & 0 & 0 & I
\end{array}
$$

Structure of one-step transition matrix, P

$$
\begin{array}{c}
 \\ M \\ \vdots \\ i \\ \vdots \\ 2
\end{array}
\begin{array}{c}
\begin{array}{ccccc}
M & & i & & 2
\end{array} \\
\left[\begin{array}{ccccc}
\frac{1-p}{M}+\frac{p}{M} & \cdots & \frac{1-p}{M}+\frac{p}{M} & \cdots & \frac{1-p}{M}+\frac{p}{M} \\
\vdots & \vdots & \vdots & \vdots & \vdots \\
0 & \cdots & \frac{1-p}{M}+\frac{p}{i}+\frac{(M-i)(1-t)(1-p)}{M} & \cdots & \frac{1-p}{M}+\frac{p}{i} \\
\vdots & \vdots & \vdots & \vdots & \vdots \\
0 & \cdots & 0 & \cdots & \frac{1-p}{M}+\frac{p}{2}+\frac{(M-2)(1-t)(1-p)}{M}
\end{array}\right]
\end{array}
$$

Structure of Q^*

$$
\begin{array}{c}
 \\ M \\ \vdots \\ i \\ \vdots \\ 2
\end{array}
\begin{array}{c}
\begin{array}{ccccc}
M' & & i+1' & & 2'
\end{array} \\
\left[\begin{array}{ccccc}
0 & \cdots & 0 & \cdots & 0 \\
\vdots & \vdots & \vdots & \vdots & \vdots \\
\frac{(1-p)t}{M} & \cdots & \frac{(1-p)t}{M} & \cdots & 0 \\
\vdots & \vdots & \vdots & \vdots & \vdots \\
\frac{(1-p)t}{M} & \cdots & \frac{(1-p)t}{M} & \cdots & 0
\end{array}\right]
\end{array}
$$

Structure of R^*

Figure 5.6. Structure of the one-step transition matrix for the embedded Markov chain model with entries for the combined $(1-p)$PRS + pPAS algorithm using a uniform distribution and acceptance probability t.

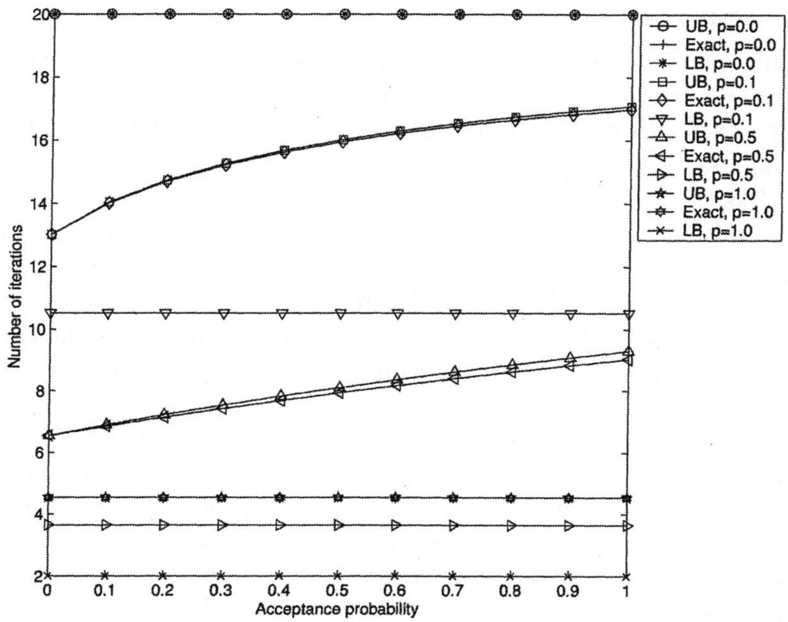

Figure 5.7. Upper bound (UB), lower bound (LB) and exact expected number of iterations to convergence for Example 1 with p=0.0, p=0.1, p=0.5 and p=1.0.

Figure 5.7 shows the upper and lower bounds on expected number of iterations of the combined PRS-PAS algorithm compared to the exact expected number of iterations for four different values, $p = 0.0, 0.1, 0.5, 1.0$ and $M = 20$. When $p = 1.0$, the algorithm is PAS; when $p = 0.0$, the algorithm is PRS. The exact expected number of iterations is calculated numerically by solving for $E[N]$ in the equation $E[N] = (I - Q)^{-1}e$ for 20 states. Notice how close the upper bound is to the exact expected number of iterations. Contrast this with Example 1 where the exact expected number of iterations was very close to the lower bound. In addition, the acceptance probability does not speed up the algorithm; on the contrary, as the acceptance probability increases the expected number of iterations also increases. That is, adding acceptance probability to the PRS-PAS algorithm seems to only degrade the performance of the algorithm, while Example 1 showed just the opposite. In the combined PRS-PAS algorithm the probability of sampling the global optimum increases or stays the same as the objective function value decreases. Therefore, there is no incentive to accept a non-improving point, that only degrade performance because the probability of sampling the global optimum is lower.

3. Summary

This chapter presents two versions of backtracking adaptive search, which allow an analysis of random search algorithms that have a probability of accepting non-improving points and consequently backtracking.

The first version of backtracking adaptive search allows a general sampling distribution which includes global optimization problems with mixed continuous and discrete variables. It is governed by its underlying sampling distribution and three probability functions: the worsening probability, hesitation probability and improving probability. A complete analysis of mixed BAS yields the distribution for the expected number of iterations to termination. The mean is given as an integral equation.

The second version of backtracking adaptive search is restricted to finite global optimization problems, but allows a general sampling probability based on a generator and acceptance probabilities. The analysis began with a Markov chain on the domain, and then developed a Markov chain on the range providing we assume the domain Markov chain is lumpable. We then developed the embedded Markov chain model which involves upper and lower triangular matrices, making the analysis tractable. We present expressions for exact solutions as well as upper and lower bounds on the expected number of iterations to first find a global optimum.

The modeling of backtracking algorithms is intended to provide insights into the performance of realistic algorithms. In a multimodal optimization problem, the benefit of moving to a non-improving point is due to a potential increased likelihood of sampling near the global optimum. The mixed BAS algorithm tries to quantify the impact of worsening through the change in sampling distributions. The discrete BAS algorithm uses a different approach by developing both domain and range models. Both approaches allow different analyses to be performed that explore the relationship between the sampling distribution and problem structure of a global optimization problem.

Chapter 6

HIT-AND-RUN BASED ALGORITHMS

The theoretical performance of stochastic adaptive search methods, as analyzed in previous chapters, is governed by the underlying sampling distribution of the global optimization algorithm. In particular, if one could sample uniformly in the improving level set of an optimization problem, as in pure adaptive search, then the expected number of iterations would be *linear* in dimension. The analyses of hesitant adaptive search and annealing adaptive search suggest that nonuniform sampling distributions may also provide polynomial performance as long as the sampling distribution has a high probability of generating an improving point. This chapter presents an approach to create a practical approximation of pure adaptive search by embedding a version of Hit-and-Run within the context of a simulated annealing algorithm. The ultimate goal is to develop an algorithm with polynomial complexity in the expected number of sample points.

Hit-and-Run is a Markov chain Monte Carlo sampler that can converge to an arbitrary target distribution, including uniform and Boltzmann distributions. Convergence properties of Hit-and-Run are summarized in Section 1. Section 2 describes Improving Hit-and-Run (IHR); an algorithm which uses Hit-and-Run to generate candidate points and accepts only those that are improving in objective function value. Although not preserving the linearity of pure adaptive search, IHR is shown in Section 2.2 to be polynomial in dimension for a class of elliptical programs. Section 3 defines Hide-and-Seek which couples the Hit-and-Run generator with an acceptance criterion and cooling schedule to approximate the performance of annealing adaptive search, and the record values approximate adaptive search. Section 4 describes several extensions to the Hit-and-Run based optimization methods, including several methods to generate directions. Up until this point the chapter is aimed at continuous global optimization problems; however Section 4.2 discusses current research on modifying Hit-and-Run to solve discrete or mixed continuous/discrete global optimization problems. Finally, Section 5 summarizes computational experience with many of the extensions.

1. Hit-and-Run

Hit-and-Run, introduced by Smith in 1984 [157], is a Markov chain Monte Carlo sampling technique which iteratively generates a sequence of points in a set by taking steps of random length in randomly generated directions. Hit-and-Run can be applied to any bounded region in $\mathbb{R}^n$, and has nice convergence properties. Smith [157] showed that Hit-and-Run can generate a sequence of points that asymptotically approach a uniform distribution on open sets of full dimension, and modifications of Hit-and-Run can approximate arbitrary multivariate distributions [16], including the Boltzmann distribution. For this reason Hit-and-Run is an appropriate sampling technique to approximate the desirable properties of pure adaptive search.

There has been much interest in Hit-and-Run as a Markov chain Monte Carlo sampler. Hit-and-Run is among the first Markov chain Monte Carlo sampler in the category of line samplers [2]. The Hit-and-Run method was originally applied to identify nonredundant constraints in linear programs [19, 25], while Caron, et al. [32] elaborates on its performance for detecting necessary (nonredundant) constraints in systems of linear constraints. Hit-and-Run has also been applied to stochastic programming [22], evaluation of multidimensional integrals [34], statistical model validation [98], and "the shopping basket" multicategory decision problem [108].

Hit-and-Run [157] generates a sequence of points $\{X_k, k = 0, 1, \ldots\}$ on a bounded and full dimensional set $S \subseteq \mathbb{R}^n$ as follows.

Hit-and-Run

Step 0. Initialize $X_0 \in S$ and set $k = 0$.

Step 1. Generate a random direction D_k uniformly distributed over a direction set $D \subseteq \mathbb{R}^n$.

Step 2. Generate a random point $X_{k+1} = X_k + \lambda D_k$ uniformly distributed over the line set

$$L_k = \{x : x \in S \text{ and } x = X_k + \lambda D_k,\ \lambda \text{ a real scalar}\}.$$

If $L_k = \emptyset$, go to Step 1.

Step 3. If a stopping criterion is met, stop. Otherwise increment k and return to Step 1.

1.1 Implementation of Hit-and-Run

The basic steps of generating a random direction and a random point on the line set are discussed. In Step 1 a random direction is generated over a specified set of directions. The most common direction set, and one that is readily implemented, is the set of directions on the unit hypersphere in $\mathbb{R}^n$. Generating a uniformly distributed direction on an n-dimensional hypersphere is termed *hyperspherical direction (HD)* in [19, 185]. It is easily implemented by generating n independent values $d_i, i = 1, 2, \ldots, n$ from a standard normal distribution, $N(0,1)$ and scaling them to determine the unit direction vector D;

$$D = (d_1, d_2, \ldots, d_n) \left(\sum_{i=1}^{n} {d_i}^2 \right)^{-1/2} \tag{6.1}$$

(see [90]).

Other direction sets are possible, such as a *coordinate direction (CD)* set where the coordinate directions are sampled uniformly. Smith [157] mentions that CD has possible computational advantages as well as disadvantages over the HD direction set. Coordinate directions and other direction sets are discussed in Section 4.1.

Generating a random point on the line set in Step 2 is straightforward as long as it is possible to determine the points where the line through X_k in the direction D_k intersects S. If S is a convex set, first determine the intersection points, that is, find λ_{min} and λ_{max} such that $X_k + \lambda D_k \in S$ for $\lambda_{min} \leq \lambda \leq \lambda_{max}$. Then λ can be chosen uniformly over that interval thus producing the random point.

This is particularly easy for the case where S is a polytope described by m linear constraints, $S = \{x : Ax \leq b\}$, and A is an $m \times n$ matrix with entries a_{ij}, b is an $m \times 1$ vector with entries b_i. Given the current point X_k and direction D_k, we need to find bounds on λ such that $A(X_k + \lambda D_k) \leq b$. This can be easily found with a modified minimum ratio test [113]. The current point X_k is an $n \times 1$ vector with entries x_j and the current direction D_k is an $n \times 1$ vector with entries d_j. The ith constraint can be written

$$\sum_{j=1}^{n} a_{ij}(x_j + \lambda d_j) \leq b_i$$

for $i = 1, \ldots, m$. This simplifies to

$$\lambda \sum_{j=1}^{n} a_{ij} d_j \leq b_i - \sum_{j=1}^{n} a_{ij} x_j$$

for $i = 1, \ldots, m$. The range on λ can be determined by

$$\begin{aligned} \lambda &\leq \lambda_{max} \\ &= \text{minimum} \left\{ \frac{b_i - \sum_{j=1}^{n} a_{ij} x_j}{\sum_{j=1}^{n} a_{ij} d_j} \text{ for } i \text{ such that } \sum_{j=1}^{n} a_{ij} d_j > 0 \right\} \\ \lambda &\geq \lambda_{min} \\ &= \text{maximum} \left\{ \frac{b_i - \sum_{j=1}^{n} a_{ij} x_j}{\sum_{j=1}^{n} a_{ij} d_j} \text{ for } i \text{ such that } \sum_{j=1}^{n} a_{ij} d_j < 0 \right\}. \end{aligned}$$

The upper bound λ_{max} is found by determining the minimum distance to an intersection with a constraint moving in one direction, and the lower bound λ_{min} is the opposite. Once these upper and lower bounds are computed, a random number λ is generated that is uniformly distributed between these limits. This gives the next point, $X_{k+1} = X_k + \lambda D_k$.

If the feasible region S is nonconvex, and/or the intersection points are not easily determined, one implementation alternative is to enclose the feasible region in a box (hyperrectangle) and use a one-dimensional acceptance-rejection along the resulting line segment until a feasible point is found. Specifically, suppose the feasible region S is contained in the following set of box constraints, i.e., for all $X = (x_1, \ldots, x_n) \in S$,

$$L_j \leq x_j \leq U_j \qquad \text{for } j = 1, \ldots, n$$

where the upper bound is greater than the lower bound, $U_j > L_j$ for $j = 1, \ldots, n$. In this case, we need to find the range on λ that satisfies

$$L_j \leq x_j + \lambda d_j \leq U_j$$

or rewriting slightly,

$$L_j - x_j \leq \lambda d_j \leq U_j - x_j$$

for $j = 1, \ldots, n$. Then upper and lower bounds on λ can be determined by

$$\begin{aligned}
\lambda &\leq \lambda_{max} \\
&= \text{minimum} \left\{ \begin{array}{ll} \dfrac{U_j - x_j}{d_j} & \text{for } j \text{ such that } d_j > 0 \\ \dfrac{L_j - x_j}{d_j} & \text{for } j \text{ such that } d_j < 0 \end{array} \right\} \\
\lambda &\geq \lambda_{min} \\
&= \text{maximum} \left\{ \begin{array}{ll} \dfrac{L_j - x_j}{d_j} & \text{for } j \text{ such that } d_j > 0 \\ \dfrac{U_j - x_j}{d_j} & \text{for } j \text{ such that } d_j < 0 \end{array} \right\}.
\end{aligned}$$

These upper and lower bounds on λ are easy to calculate, and then λ is generated uniformly within this interval. If the resulting point is feasible, $X_{k+1} = X_k + \lambda D_k \in S$, then the step is complete. If not, generate another λ until a feasible point is found.

The efficiency of this one-dimensional acceptance-rejection procedure depends on how tightly the bounding set contains S. A stalling problem may occur if the line intersects a small portion of S [85]. For example, in Figure 6.1 the current point is near a boundary of S, and the line passing through X_k has a small intersection with S. For the direction illustrated, the algorithm is likely to sample many λ's before finding a feasible point X_{k+1}. Computational experience indicates [67] that limiting the number of sample points to one per direction performed better than allowing multiple sample points (e.g. λ's) before resampling a direction. This empirical finding supports the idea of choosing a random direction and one random point. If the candidate point is infeasible or non-improving, then reject both the point and direction, and generate another random direction. The problem of stalling can also be addressed by considering alternate sets of directions (see Section 4.1).

1.2 Convergence to Uniform Distribution

Hit-and-Run is known to generate points that are asymptotically uniformly distributed on an open full-dimensional set [157]. This means that the points generated by Hit-and-Run eventually exhibit frequencies in accordance with a uniform distribution. This convergence result relies on properties of the Markov chain of points produced by Hit-and-Run.

The one-step transition probability of Hit-and-Run determines its behavior. The transition probability of sampling a point within a measurable set $A \subseteq S$ depends on the probability density $f(y|x)$ of y given x,

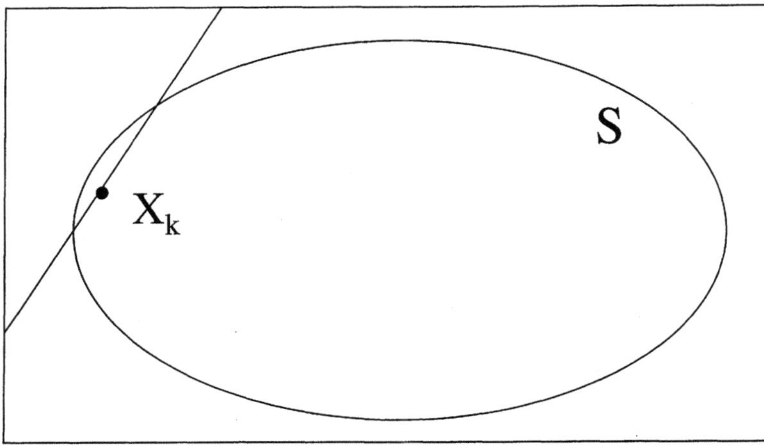

Figure 6.1. Hit-and-Run may stall when the line intersects a small portion of S.

and can be characterized by

$$P\left(X_{k+1} \in A | X_k = x\right) = \int_A f(y|x)dy$$

and for the hyperspherical direction set,

$$= \int_{\partial D} \frac{\|L_A(d,x)\|}{\|L_S(d,x)\|} dF_{\partial D}(d) \qquad (6.2)$$

where ∂D is the boundary of the unit sphere, and $F_{\partial D}(\cdot)$ is the cumulative distribution function for the uniformly distributed direction vector on the boundary of the unit sphere. The expression $\|L_A(d,x)\|$ is the combined lengths of the line segments formed by the intersection of the set A with the line in direction d originating at x. In words, the probability of sampling a point in A, starting at point x, is the probability of landing within A given direction d, i.e. $\|L_A(d,x)\|/\|L_S(d,x)\|$, integrated over all possible directions.

In general, it is difficult to provide an exact expression for the one-step distribution for Hit-and-Run, however it is possible to show that the Markov chain satisfies two specific properties that are important to show convergence to the uniform distribution. The first property is that the transition probability density function for Hit-and-Run is symmetric, i.e., $f(y|x) = f(x|y)$ for all $x, y \in S$. Intuitively, the probability of going from x to y is the same as the probability of going from y to x because they share the same line connecting the two points, and the sampling density along that line does not depend on whether it is coming from x or y [157]. The second property is that, with a single step, Hit-and-Run has a positive probability of sampling anywhere in the space. This gives Hit-and-Run the feature of global sampling, which is in contrast to many sampling techniques that are limited to sampling within local neighborhoods of the current point. These two properties of being symmetric and having a positive probability of sampling anywhere in the set are enough to show convergence to the uniform distribution regardless of the starting point. The result, first stated in [157, Theorem 2], is given below.

THEOREM 6.1 *The Markov chain $X_0, X_1, X_2, \ldots$ induced by the Hit-and-Run algorithm over an open bounded set $S \subset \mathbb{R}^n$ converges to the uniform distribution for all starting points $x \in S$ and all measurable sets $A \subseteq S$, i.e.,*

$$\lim_{m \to \infty} P(X_m \in A | X_0 = x) = \nu(A)$$

where ν is the uniform distribution.

Proof: See [157]. ■

While Hit-and-Run converges to a uniform distribution, the one-step distribution has a local bias. To see this effect, consider an n-dimensional unit ball, and execute one iteration of Hit-and-Run from a point in the center. Figure 6.2 shows an annulus at radius r_1 and an annulus at radius r_2, of the same thickness dr. Hit-and-Run gives equal probability to each annulus, because the ratio of the line segments is the same in both cases. Let $f(r)$ be the sampling density at a distance r from the center, and let $S(r)$ be the surface area of the n-dimensional hypersphere of radius r. Since Hit-and-Run gives equal probability to each annulus, we have

$$f(r_1)S(r_1)dr = f(r_2)S(r_2)dr$$

and hence,

$$\frac{f(r_1)}{f(r_2)} = \frac{S(r_2)}{S(r_1)} = \left(\frac{r_2}{r_1}\right)^{n-1}$$

which implies that it gets harder to sample a point far away from the current point as the dimension n increases (see [93, page 68 and Appendix 1] and [190]). A truly uniform search on the hypersphere inverts the density ratio found for Hit-and-Run.

A simulation of one thousand points generated from the center of a square in two dimensions is presented in the top portion of Figure 6.3. The points demonstrate the one-step transition density of Hit-and-Run. The clustering of points around the center illustrate that the one-step transition probabilities have a local bias. Nevertheless, the theory says

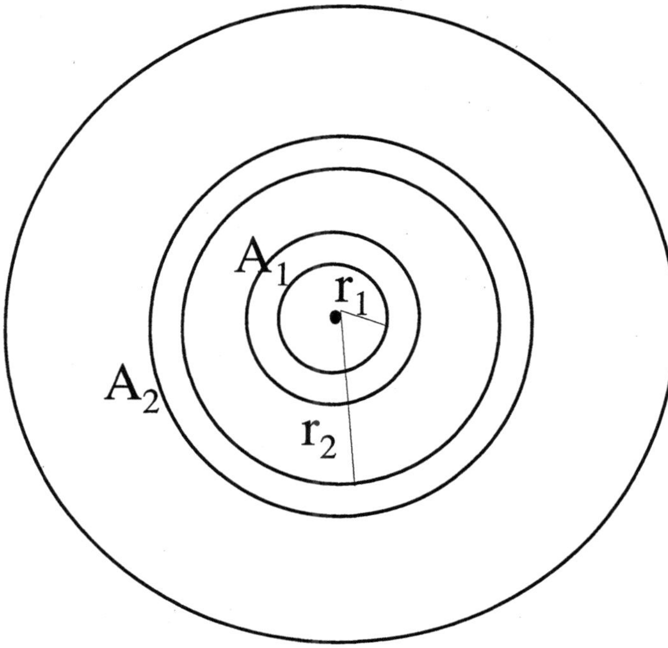

Figure 6.2. Generating a point from the center of a hypersphere using Hit-and-Run.

that eventually the sequence of points generated by Hit-and-Run converges to a uniform distribution. The bottom portion of Figure 6.3 presents a simulation of a sequence of one thousand iterations of Hit-and-Run performed on the same two dimensional square. The initial point is in the center of the square. Visually this simulation illustrates that the points resemble a uniform distribution.

1.3 Metropolis Hit-and-Run

Hit-and-Run can also be used to emulate other distributions besides the uniform distribution. Hit-and-Run was extended by adding an acceptance criterion, so providing a Markov chain Monte Carlo sampling method that converges to an arbitrary multivariate distribution [16, 140]. The idea is to use the same technique to generate a random direction and random point on the line, and then accept the candidate point with a probability that reflects the target distribution. This acceptance probability is sometimes called the Metropolis criterion, and can be viewed as a Metropolis filter - hence the name Metropolis Hit-and-Run [85].

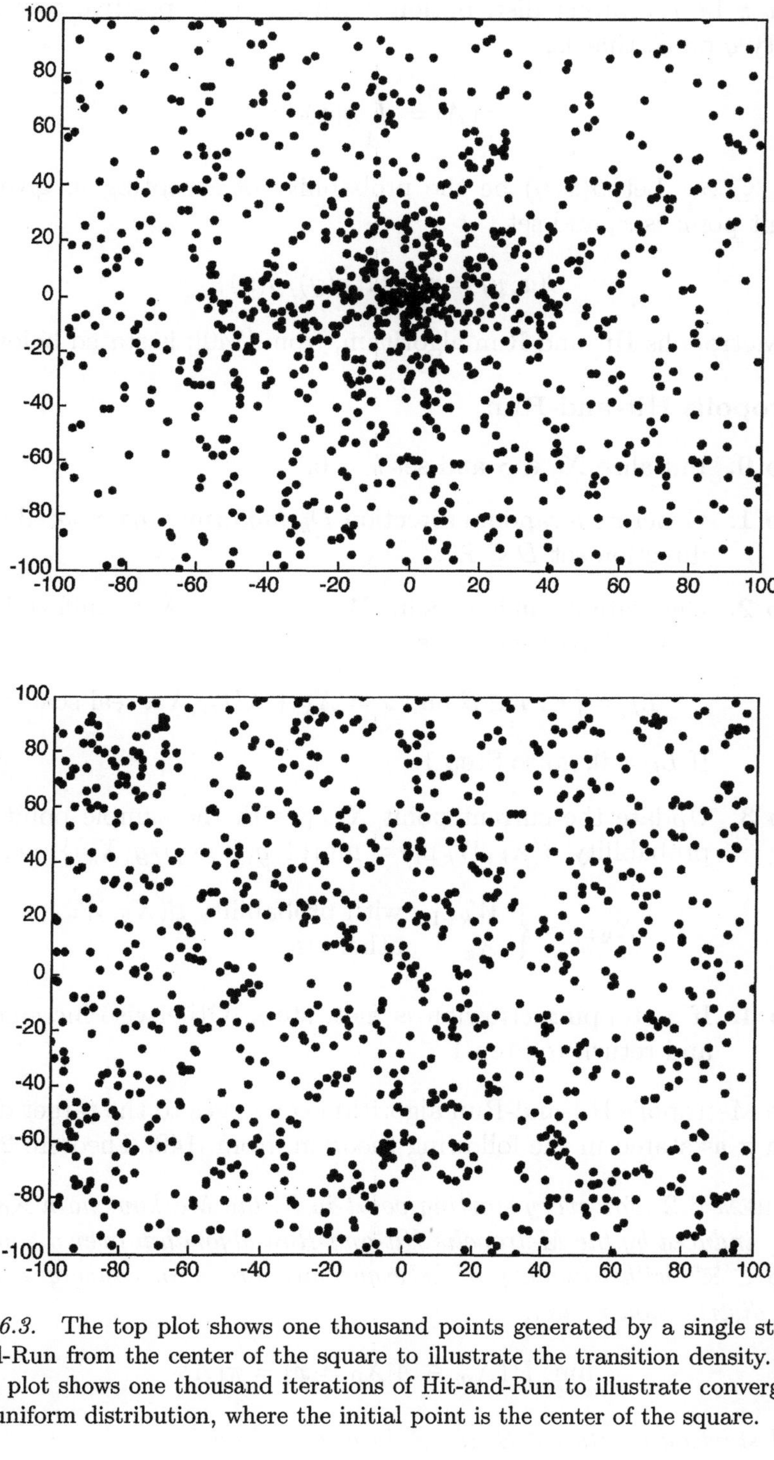

Figure 6.3. The top plot shows one thousand points generated by a single step of Hit-and-Run from the center of the square to illustrate the transition density. The bottom plot shows one thousand iterations of Hit-and-Run to illustrate convergence to the uniform distribution, where the initial point is the center of the square.

Let π be the target distribution, with a strictly positive continuous density g on S, that is,

$$\pi(A) = \int_A g(x)dx$$

for $A \subseteq S$. Let $\beta(x,y)$ be the probability of accepting y, given the current point is x, and set

$$\beta(x,y) = \min\{1, g(y)/g(x)\}.$$

The Metropolis Hit-and-Run algorithm, from [140], is stated below.

Metropolis Hit-and-Run

Step 0. Initialize $X_0 \in S$ and set $k = 0$.

Step 1. Generate a random direction D_k uniformly distributed over a direction set $D \subseteq \mathbb{R}^n$.

Step 2. Generate a random point $W_{k+1} = X_k + \lambda D_k$ uniformly distributed over the line set

$$L_k = \{x : x \in S \text{ and } x = X_k + \lambda D_k, \;\; \lambda \text{ a real scalar}\}.$$

If $L_k = \emptyset$, go to Step 1.

Step 3. Update the current point X_{k+1} with the sample point with probability $\beta(X_k, W_{k+1}) = \min\{1, g(W_{k+1})/g(X_k)\}$, i.e. set

$$X_{k+1} = \begin{cases} W_{k+1} & \text{with probability } \beta(X_k, W_{k+1}) \\ X_k & \text{otherwise.} \end{cases}$$

Step 4. If a stopping criterion is met, stop. Otherwise increment k and return to Step 1.

The Metropolis Hit-and-Run algorithm converges to the target distribution π as stated in the following theorem, from [140, Theorem 2.5].

THEOREM 6.2 *For every starting point in S, the Markov chain $X_0, X_1, X_2, \ldots$ induced by the Metropolis Hit-and-Run algorithm over a compact body $S \subset \mathbb{R}^n$ with boundary of Lebesgue measure zero, converges to the target distribution π, i.e.,*

$$\lim_{m \to \infty} P(X_m \in A | X_0 = x) = \pi(A)$$

for all starting points $x \in S$ and A in a σ-field on S.

Proof: See [140]. ■

The theorem of convergence to a target distribution allows a broad range of acceptance probabilities. Three conditions on the acceptance probability β are: i) it is uniformly bounded away from zero, ii) jointly measurable, and iii) satisfies the symmetry property,

$$\beta(x,y)g(x) = \beta(y,x)g(y) \quad \text{for all } x,y \in S$$

as given in [140]. For our purposes, we are primarily interested in using the Boltzmann distribution as the target π, and thus $\beta(x,y) = \min\{1, g(y)/g(x)\}$ where $g(x)$ is set to the Boltzmann density. In this case, the acceptance probability of Metropolis Hit-and-Run is the acceptance probability used in simulated annealing algorithms. When $\beta(x,y) = 1$, all candidate points are accepted and the random walk is the Hit-and-Run algorithm, so iterates are asymptotically uniformly distributed. Choosing the target distribution as a Boltzmann distribution and embedding it in a global optimization algorithm is used in Section 3 to approximate annealing adaptive search.

1.4 Rate of Convergence to Target Distribution

A practical matter of interest is how fast Hit-and-Run converges to the uniform distribution, or to a more general target distribution. Smith presented the first bound on convergence rate of Hit-and-Run in [157, Theorem 3].

THEOREM 6.3 *Let $X_0, X_1, X_2, \ldots$ be the Markov chain generated by the Hit-and-Run algorithm over an open bounded set $S \subset \mathbb{R}^n$. Then for any measurable set $A \subseteq S$,*

$$|P(X_m \in A|X_0 = x) - \nu(A)| < (1 - (\gamma/n2^{n-1}))^{m-1}$$

where ν is the uniform distribution and γ is the ratio of the n-dimensional content of S to the n-dimensional content of the smallest sphere containing S.

Proof: See [157]. ■

As a numerical illustration of this bound on the rate of convergence, consider the case when S is a disk in $\mathbb{R}^2$. Then

$$|P(X_m \in A|X_0 = x) - \mu(A)| < (3/4)^{m-1}$$

which is exponentially fast. Refer again to Figure 6.3 which plots one thousand iterations of Hit-and-Run in a square in two dimensions. For this example,

$$|P(X_{1000} \in A|X_0 = (0,0)) - \nu(A)| < 10^{-75}$$

where the bound is calculated using $\gamma = (200)^2/4\pi\sqrt{20,000}^2$, $n = 2$, and $m = 1,000$. The bound on convergence rate given in Theorem 6.3 is only of interest in low dimensions because the bound can overestimate the number of iterations by several orders of magnitude [157].

Lovász, in 1999 [102], derived tighter bounds on the rate of convergence for Hit-and-Run. He proved that Hit-and-Run on a convex body in n dimensions produces an approximately uniformly distributed sample point in polynomial time, $O(n^3)$, which matches the best known bound for such a sampling algorithm.

While Hit-and-Run can be made to converge to any target distribution, in general the convergence rate may be slow. Bélisle showed, in 2000 [15], that the convergence rate of Hit-and-Run can be severely affected by small perturbations in the target distribution.

The characteristics of Hit-and-Run suggest that Hit-and-Run can be used in a global optimization algorithm. The approach is to execute Hit-and-Run for some number of iterations, such as proportional to n^3, for each level set in order to generate an asymptotically uniformly distributed point in the level set, and then pure adaptive search predicts that on the order of n such iterations would be needed. The number of iterations of Hit-and-Run for each level set is not precisely determined, but the concept motivates the use of Hit-and-Run in an optimization context.

2. Improving Hit-and-Run (IHR)

The main idea behind Improving Hit-and-Run is to embed Hit-and-Run as a method of approximating uniformly distributed points within an iterative optimization procedure, to attempt to approximate the linear performance of pure adaptive search. Improving Hit-and-Run (IHR), introduced by Zabinsky et al. [189] was the first time Hit-and-Run was applied to a global optimization problem. The term "improving" was coined to indicate that the sequence of points were improving with regard to their objective function values as in pure adaptive search.

To motivate IHR, one way to approximate PAS is to use Hit-and-Run to generate approximately uniform points, then select those that land in improving level sets. If the uniform approximation was exact, then only a linear number of such improving iterations are needed, on average. Since a sequence of Hit-and-Run is only asymptotically guaranteed to be uniform within the improving region, we must decide on how long a sequence to generate. At one extreme, when the Hit-and-Run sequences are very long and provide a close approximation to sampling from a uniform distribution, we might expect a good approximation to pure adaptive search and a nearly linear number of improving points in dimension. Unfortunately this implies a linear number of nearly infinite sequences: an impractical approach. At the other extreme, the Hit-and-Run sequence could be reduced to a length of *one*. Although the sequence of points generated per iteration is insufficiently long to closely approximate uniformity, the hope is that the algorithm may be similar enough to pure adaptive search to inherit a polynomial complexity. This is the motivation for IHR, which uses Hit-and-Run sequences of length one. In fact, for the class of positive definite quadratic programs, the expected number of *function evaluations* for IHR remains polynomial, in particular, $O(n^{5/2})$ [189]. The following section formally defines Improving Hit-and-Run, and Section 2.2 presents the complexity analysis.

2.1 Definition of Improving Hit-and-Run

Improving Hit-and-Run is defined on the global optimization problem,

$$(P) \qquad \min_{x \in S} f(x)$$

where x is an n-dimensional vector, the feasible region S is a convex, compact, full-dimensional subset of $\mathbb{R}^n$, whose boundary has Lebesgue measure zero, and f is a real-valued continuous function defined over S.

The basic structure of IHR is to generate a random direction followed by a candidate point that is uniformly distributed along that direction. The generation of the candidate point uses Hit-and-Run, and then the candidate point is accepted only if it is improving in objective function value. IHR can also be viewed as a sequential step size algorithm (Chapter 1, Section 4.2) with a random direction D_k and a random step size λ. A positive definite matrix H controls the direction distribution and typically H is taken to be the identity matrix. In Section 2.2, H is taken to be the Hessian of a quadratic objective function.

The Improving Hit-and-Run algorithm [189] is defined as follows.

Improving Hit-and-Run (IHR)

Step 0. Initialize $X_0 \in S$, $Y_0 = f(X_0)$, and set $k = 0$.

Step 1. Generate a random direction vector D_k from the multivariate normal distribution with mean 0 and covariance matrix H^{-1}. If $H = I$, this is equivalent to generating a direction uniformly distributed on the boundary of a unit hypersphere.

Step 2. Generate a candidate point $W_{k+1} = X_k + \lambda D_k$ by sampling uniformly over the line set

$$L_k = \{x : x \in S \text{ and } x = X_k + \lambda D_k, \lambda \text{ a real scalar}\}.$$

If $L_k = \emptyset$, go to Step 1.

Step 3. Update the current point X_{k+1} with the candidate point if it is improving, i.e. set

$$X_{k+1} = \begin{cases} W_{k+1} & \text{if } f(X_k + \lambda D_k) < Y_k \\ X_k & \text{otherwise} \end{cases}$$

and set $Y_{k+1} = f(X_{k+1})$.

Step 4. If the stopping criterion is met, stop. Otherwise increment k and return to Step 1.

Possible choices for each step in the algorithm are discussed next.

To begin in Step 0, Improving Hit-and-Run needs a feasible point. If an initial point is not readily available, one approach is to define upper and lower bounds on x that contain the feasible region. Specifically, define box constraints such that S is contained in the box B where $S \subseteq B = \{x \in \mathbb{R}^n : l_i \leq x_i \leq u_i, i = 1, \ldots, n\}$. Then a Phase I approach can be taken (as in linear programming [113]), applying the algorithm to the problem with box constraints, and minimizing the sum of infeasibilities. For instance, if S is defined by the intersection of non-linear inequalities, $S = \{x \in \mathbb{R}^n : g_j(x) \leq 0, j = 1, \ldots, m\}$ and that S is contained in the box B, then the Phase I problem can be written $\min_{x \in B} \sum_{j=1}^{m} \max\{0, g_j(x)\}$. Any feasible solution in S will have an objective function value of zero in this mini-max Phase 1 problem. Alternatively, if an initial feasible point is not available, a penalty function approach can be taken (see Chapter 7, Section 1.2) to create a modified problem with a penalty factor for the constraints.

Generating a random direction vector, as defined in Step 1, first requires establishing a positive definite matrix H. If the matrix H is the identity matrix, then the direction distribution is uniform on a hypersphere and can be implemented using Equation 6.1. This is an easy direction distribution to implement. In practice, for a global optimization problem H might be locally estimated as in quasi-Newton local search procedures. The choice of the direction distribution controlled by H is discussed in more detail in Section 4.1.

To generate a random direction vector for a general choice of H, first find a matrix A of rank n such that $H = A'A$. Then generate a random direction vector D_k uniform on a hypersphere, as in Equation 6.1, and transform it with $A^{-1}D_k$. Then $A^{-1}D_k$ has the desired distribution, i.e. $A^{-1}D_k$ is normally distributed with mean 0 and covariance matrix H^{-1}.

There are several ways to obtain a matrix A such that $H = A'A$. One possibility is to perform a Cholesky decomposition $H = LL'$, where L is a lower triangular matrix, yielding $A = L'$. Another possibility is to diagonalize H, by writing $H = ZDZ'$ where D is a diagonal matrix containing the eigenvalues of H, and Z is a matrix having the corresponding eigenvectors as its columns. This approach yields $A = ZD^{\frac{1}{2}}Z'$.

Step 2 requires generating a point uniformly distributed over a line set. The length of the line segment must be determined, as well as the points of intersection of the direction and the feasible set to generate λ_k. If S is an n-dimensional polytope described by linear inequalities (e.g. box constraints), this is easily computed by using a slight modification of the minimum ratio test. The details for calculating λ_{min} and λ_{max} are given in Section 1.1. If S is a convex set defined as the intersection of functional inequalities, e.g. $S = \{x \in \mathbb{R}^n : g_j(x) \leq 0, j = 1, \ldots, m\}$, then the intersection points can be determined by taking the inverse of the constraint equations. However, if the constraint equations are not easily invertible, or if the feasible region is not convex, than a common approach is to define box constraints that contain the feasible region, and use an acceptance/rejection scheme to generate the candidate point along the direction. As mentioned in Section 1.1, computational experience using a one-dimensional acceptance-rejection procedure indicates that it is more efficient to sample once per direction and if that point is not feasible, return to Step 1 to sample a new direction, than to sample multiple times per direction [67].

Step 3 is straightforward to implement. Step 4 requires a stopping criterion, which is not addressed here. See [23, pages 853-864] for a survey of stopping rules for stochastic algorithms. Typically a maximum number of iterations or a limit on CPU time is set as a stopping criterion.

2.2 Polynomial Performance of IHR

We now turn to the analysis of the expected number of function evaluations required by IHR to achieve a specific level of accuracy. The detailed analysis is developed for a "spherical program" and the results are extended to a class of mathematical programs called "elliptical programs." The performance of IHR on an elliptical program using H in the direction distribution equals the performance of IHR on a transformed spherical program using I in the direction distribution, as shown in Theorem 6.4. Most of the analysis is then performed on the spherical program which is easier to visualize. An intermediate result in the proof of polynomial performance characterizes the conditional probability of improvement for the next sample point for IHR (in Lemma 6.7). The final development combines the conditional probability for IHR with the linearity result of PAS. The main result of $O(n^{5/2})$ complexity for IHR is stated in Corollary 6.9.

The complexity analysis holds for a class of mathematical programs with "elliptical" level sets. This class of programs includes positive definite quadratic programs as a special case.

DEFINITION 6.1 *An* elliptical program *is a mathematical program*

$$(EP) \qquad \min_{x \in S} f(x) \tag{6.3}$$

where $f(x)$ can be expressed as

$$f(x) = h(r) \qquad \text{with} \qquad r = \|x - x_*\|_A,$$

where A is an $n \times n$-matrix of full rank, and h is strictly monotonically increasing for $r \geq 0$. The norm $\|\cdot\|_A$ is defined for any n-dimensional point z as

$$\|z\|_A = \|Az\|$$

where $\|\cdot\|$ denotes the standard Euclidean norm. Note that $\|\cdot\|_I = \|\cdot\|$. The feasible set S for an elliptical program is taken as the level set associated with X_0, so that $Y_0 = y^*$.

A spherical program *(SP) is a special case of an elliptical program when A is the identity matrix.*

When the IHR algorithm is applied to an elliptical program, matrix H used in the direction distribution is chosen to be equal to $A'A$. In the special case of a spherical program, IHR uses $A = I$ in the direction distribution resulting in a hyperspherical direction.

An elliptical program can be interpreted geometrically as a problem with level sets of f that are elliptical in shape, and nested about the optimum. The function $h(r)$ can be interpreted as a means of layering the level sets. In the special case where $A = I$ the level sets become spherical in shape, hence the name spherical program, and the variable r can be geometrically interpreted as the radius of the level set. A positive definite quadratic program falls into the class of elliptical programs. Notice that an elliptical program is convex if and only if $h(r)$ is convex in r. The class of elliptical programs also includes cases where $h(r)$ is nonconvex in r. Figure 6.4 illustrates a spherical program in two variables, where the level sets are concentric circles. The figure also illustrates an elliptical program in two variables where the level sets are concentric ellipses, and the problem is unimodal, but not convex.

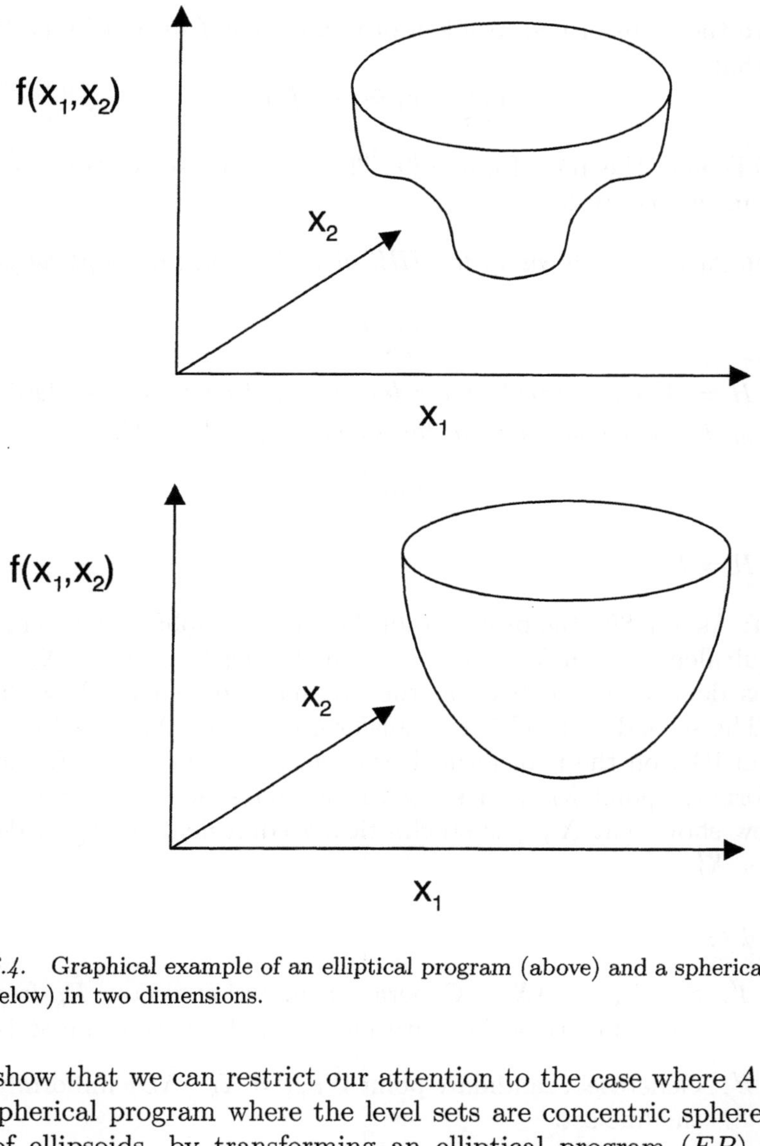

Figure 6.4. Graphical example of an elliptical program (above) and a spherical program (below) in two dimensions.

We show that we can restrict our attention to the case where $A = I$, i.e. a spherical program where the level sets are concentric spheres instead of ellipsoids, by transforming an elliptical program (EP) with corresponding matrix A into a spherical program $(\widetilde{EP})$. The transformation defines a transformed point $\tilde{x} = Ax$, and

$$\tilde{S} = \{\tilde{x} \in \mathbb{R}^n : A^{-1}\tilde{x} \in S\}$$

and

$$\tilde{f} : \tilde{S} \to \mathbb{R}$$
$$\tilde{f}(\tilde{x}) = f(A^{-1}\tilde{x}).$$

Denote the transformed problem of minimizing $\tilde{f}$ over $\tilde{S}$ by $(\widetilde{EP})$, and note that

$$\min_{x \in S} f(x) = \min_{\tilde{x} \in \tilde{S}} \tilde{f}(\tilde{x}).$$

The following theorem, from [189, Theorem 3.1], states that these two problems are equivalent.

THEOREM 6.4 *Performing the IHR algorithm on the elliptical problem* (EP)

$$\min_{x \in S} f(x)$$

using $H = A'A$ *is equivalent (under the identification* $\tilde{x} \leftrightarrow Ax$*) to performing the IHR algorithm on the spherical problem* $(\widetilde{EP})$

$$\min_{\tilde{x} \in \tilde{S}} \tilde{f}(\tilde{x})$$

using $H = I$.

Proof: As in [189], the proof establishes that the following two methods are equivalent. Given $X_k \in S$, the first method is to generate X_{k+1} using IHR as defined in the text on the original problem (EP) with $H = A'A$. The second method is to transform the given X_k into $\tilde{X}_k = AX_k$, perform IHR on the transformed problem $(\widetilde{EP})$ with $H = I$, and then transform the point $\tilde{X}_{k+1}$ back into the original space, $X'_{k+1} = A^{-1}\tilde{X}_{k+1}$. We now show that X_{k+1} is stochastically equivalent to X'_{k+1}, denoted $X_{k+1} \sim X'_{k+1}$.

Method 2:

Step 1′. Set $\tilde{X}_k = AX_k$. Generate a direction vector $\tilde{D}_k$ from the normal distribution with mean 0 and covariance matrix I.

Step 2′. Generate a candidate point $\tilde{W}_{k+1} = \tilde{X}_k + \tilde{\lambda}\tilde{D}_k$ uniformly from the line set,

$$\tilde{L}_k = \left\{\tilde{x} : \tilde{x} \in \tilde{S} \text{ and } \tilde{x} = \tilde{X}_k + \tilde{\lambda}\tilde{D}_k, \tilde{\lambda} \in \mathbb{R}\right\}.$$

If $\tilde{L}_k = \emptyset$, go to Step 1′.

Step 3′. Update the current point X'_{k+1} with the candidate point if it is improving,

$$\tilde{X}_{k+1} = \begin{cases} \tilde{W}_{k+1} & \text{if } \tilde{f}(\tilde{X}_k + \tilde{\lambda}_k\tilde{D}_k) < \tilde{f}(\tilde{X}_k) \\ \tilde{X}_k & \text{otherwise} \end{cases}$$

and set $\tilde{Y}_{k+1} = \tilde{f}(\tilde{X}_{k+1})$. Set $X'_{k+1} = A^{-1}\tilde{X}_{k+1}$.

First note that AD_k, where D_k is from Step 1 of IHR, has the same distribution as $\tilde{D}_k$ in Step 1′ of Method 2. Using this, the set of possible step sizes, Λ_k and $\tilde{\Lambda}_k$, for both methods are the same,

$$\begin{aligned}
\Lambda_k &= \{\lambda \in \mathbb{R} : X_k + \lambda D_k \in S\} \\
&= \left\{\lambda \in \mathbb{R} : A^{-1}\tilde{X}_k + \lambda A^{-1}\tilde{D}_k \in S\right\} \\
&= \left\{\lambda \in \mathbb{R} : A^{-1}(\tilde{X}_k + \lambda \tilde{D}_k) \in S\right\} \\
&= \left\{\lambda \in \mathbb{R} : \tilde{X}_k + \lambda \tilde{D}_k \in \tilde{S}\right\} \\
&= \tilde{\Lambda}_k.
\end{aligned}$$

Thus the step size λ in Step 2 is generated from the same distribution as the step size $\tilde{\lambda}$ in Step 2′. Consequently, the candidate points from Step 2 and Step 2′ are stochastically equivalent, i.e., $W_{k+1} = X_k + \lambda D_k \sim \tilde{X}_k + \tilde{\lambda}\tilde{D}_k = \tilde{W}_{k+1}$.

Note that

$$\begin{aligned}
f(X_k + \lambda D_k) &\sim f(A^{-1}\tilde{X}_k + \tilde{\lambda} A^{-1}\tilde{D}_k) \\
&= f(A^{-1}(\tilde{X}_k + \tilde{\lambda}\tilde{D}_k)) \\
&= \tilde{f}(\tilde{X}_k + \tilde{\lambda}\tilde{D}_k)
\end{aligned}$$

and hence $\tilde{f}(\tilde{X}_{k+1}) \sim f(X_{k+1})$. Thus the probability of not finding an improvement is the same for both methods, and so if the new point is not improving, then $X'_{k+1} \sim X_{k+1}$. Now look at the distributions of the new iteration points given that an improvement occurs:

$$\begin{aligned}
X'_{k+1} &= A^{-1}\tilde{X}_{k+1} \\
&= A^{-1}(\tilde{X}_k + \tilde{\lambda}\tilde{D}_k) \\
&= A^{-1}\tilde{X}_k + \tilde{\lambda} A^{-1}\tilde{D}_k \\
&\sim X_k + \lambda D_k \\
&= X_{k+1}
\end{aligned}$$

and again $X'_{k+1} \sim X_{k+1}$. Thus performing IHR on (EP) with $H = A'A$ is equivalent in distribution to the second method of performing IHR on $(\widetilde{EP})$ with $H = I$. ■

It follows from Theorem 6.4 that in the remainder of the complexity analysis, we need only to consider the class of spherical programs. Since the number of points generated is invariant, all complexity results that are obtained for spherical programs also hold for the more general case of elliptical programs.

We next determine the conditional distribution of improvement for Improving Hit-and-Run on a spherical program. The conditional probability of making a specified improvement on a single iteration depends on the position of the current point,

$$\begin{aligned} P\left(Y_{k+1} < y | Y_k = w\right) &= E\left[P\left(Y_{k+1} < y | X_k, Y_k = w\right)\right] \\ &= E\left[P\left(X_{k+1} \in S(y) | X_k, Y_k = w\right)\right] \end{aligned}$$

where $S(y)$ is the improving set $\{x : x \in S \text{ and } f(x) < y\}$. The probability within the last expectation can be expressed as an integral as in Equation 6.2, where

$$P\left(X_{k+1} \in S(y) | X_k = x, Y_k = w\right) = \int_{\partial D} \frac{\|L_{S(y)}(d,x)\|}{\|L_S(d,x)\|} \, dF_{\partial D}(d). \quad (6.4)$$

As before, ∂D is the boundary of the unit sphere, $F_D(\cdot)$ is the cumulative distribution function for the normalized random direction vector, and $\|L_{S(y)}(d,x)\|$ is the combined lengths of the line segments formed by the intersection of the level set $S(y)$ with the line in direction d originating at x.

For general mathematical programs, the conditional probability of improvement depends on the exact location of X_k, and makes it difficult to derive a general expression. However, for the class of spherical programs, we can analytically derive the conditional probability of improvement, as stated in [189, Lemma 3.2].

LEMMA 6.5 *For any spherical program (SP), the conditional probability of improvement on the next sample point for IHR is given by*

$$\begin{aligned} & P\left(Y_{k+1} < h(s) | Y_k = h(r), Y_0 = h(q)\right) \\ & \equiv p(s; r, q) \\ & = \left(\frac{r}{q}\right) \left(\frac{s}{r}\right)^n \frac{1}{n} F_1 \left(\frac{n-1}{2}, \frac{1}{2}, \frac{1}{2}, \frac{n+2}{2}; \frac{s^2}{q^2}, \frac{s^2}{r^2}\right). \end{aligned}$$

for $0 < s \le r \le q$, *and for* $k = 0, 1, 2, \ldots$, *where* $F_1(a, b, c, d; x, y)$ *is a generalized hypergeometric function (see [5] or [57]).*

Proof: As in [189], we derive the analytical expression for $P\{Y_{k+1} < h(s) | Y_k = h(r), Y_0 = h(q)\}$ in terms of the radii of the level sets, r. In order to take full advantage of the symmetry of the problem, we use a spherical coordinate system $(\rho, \theta, \phi_i, i = 1, \ldots, n-2)$ to compute the integral required to evaluate $P\left(Y_{k+1} < h(s) | Y_k = h(r), Y_0 = h(q)\right) \equiv$

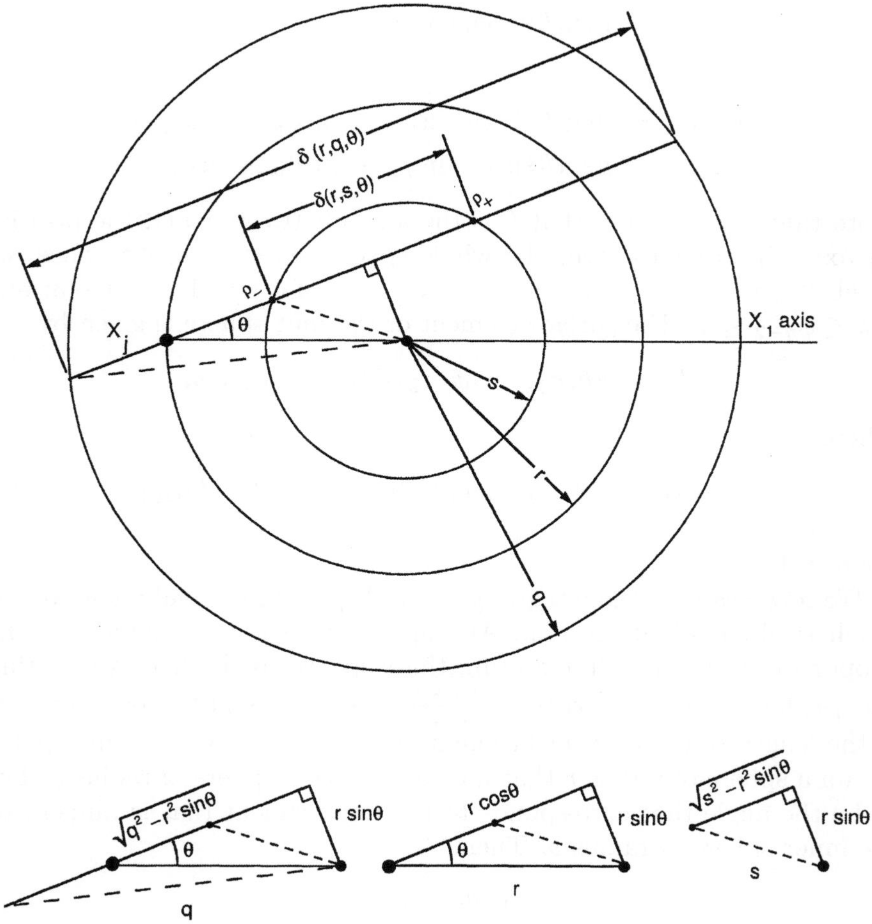

Figure 6.5. Illustration of nested level sets with notation used in the proof of Lemma 3.

$p(s; r, q)$. Let the origin of the spherical coordinates be at X_k, where the radius of the corresponding level set is r. Let the positive x_1 axis run through the center of the level sets. Figure 6.5 illustrates a cross section corresponding to fixed $\phi_i, i = 1, \ldots, n-2$ of a level set of radius r within the feasible region of radius q and containing a level set of radius s.

For a brief reminder of spherical coordinates in n dimensions, see [160, pages 227-228]. As needed in this proof, let $x_1, x_2, \ldots, x_n$ denote the Cartesian coordinates and let $\rho, \theta, \phi_1, \ldots, \phi_{n-2}$ denote the spherical coordinates. For $n \geq 3$, the relationships are given by

$$\begin{aligned} x_1 &= \rho \cos\theta \\ x_2 &= \rho \sin\theta \cos\phi_1 \end{aligned}$$

$$\begin{aligned} x_3 &= \rho \sin\theta \sin\phi_1 \cos\phi_2 \\ &\vdots \\ x_{n-1} &= \rho \sin\theta \sin\phi_1 \sin\phi_2 \cdots \sin\phi_{n-3} \cos\phi_{n-2} \\ x_n &= \rho \sin\theta \sin\phi_1 \sin\phi_2 \cdots \sin\phi_{n-3} \sin\phi_{n-2}. \end{aligned}$$

Note that $\rho = \|x\|$ and that θ is the angle between x and the positive x_1 axis. In order to cover the whole space $-\infty < x_1 < +\infty$ once, we need $0 \le r < +\infty$; $0 \le \theta \le \pi$; $0 \le \phi_i \le \pi$ for $i = 1, \ldots, n-3$; and $-\pi \le \phi_{n-2} \le \pi$. The surface element of the unit sphere is given by

$$d\omega = \omega(\theta, \phi_1, \ldots, \phi_{n-2})\, d\theta\, d\phi_1 \cdots d\phi_{n-2},$$

where,

$$\omega(\theta, \phi_1, \ldots, \phi_{n-2}) = \sin^{n-2}(\theta) \prod_{k=1}^{n-3} \sin^{n-2-k}(\phi_k)$$

for $n \ge 3$.

We now evaluate the integral given in Equation 6.4, only now we use a spherical coordinate system. An important term in the integral is the proportion of the line intersecting the s-sphere to the line intersecting the q-sphere, which we write as $\frac{\delta(r,s,\theta)}{\delta(r,q,\theta)}$ (see Figure 6.5). Here, $\delta(r, s, \theta)$ is the length of a line segment emanating in a direction θ from a point X_k on a sphere of radius r that intersects with a sphere of radius s. Let θ_0 be the angle that corresponds to the line segment that is tangent to the inner sphere of radius s. Then,

$$\sin\theta_0 = s/r.$$

Using spherical coordinates, the equation of the inner sphere of radius s is given by

$$\rho^2 - 2r\rho\cos\theta + r^2 = s^2,$$

or (refer to Figure 6.5),

$$\rho_\pm = r\cos\theta \pm \sqrt{s^2 - r^2\sin^2\theta}$$

for $0 \le \theta \le \theta_0$. Now, in the range $0 \le \theta \le \theta_0$ the length of the line segment that lies in the sphere of radius s is independent of the ϕ_i because of symmetry, and can be expressed as

$$\delta(r, s, \theta) = \rho_+ - \rho_- = 2\sqrt{s^2 - r^2\sin^2\theta}.$$

Similarly we get

$$\delta(r, q, \theta) = 2\sqrt{q^2 - r^2\sin^2\theta}.$$

The conditional probability of improvement now follows easily using the spherical coordinate system,

$$p(s;r,q) = \frac{2\int_0^{\theta_0}\int_0^{\pi}\cdots\int_0^{\pi}\int_0^{2\pi}\omega(\theta,\phi_1,\ldots,\phi_{n-2})\,\frac{\delta(r,s,\theta)}{\delta(r,q,\theta)}\,d\theta\,d\phi_1\ldots d\phi_{n-2}}{2\int_0^{\pi/2}\int_0^{\pi}\cdots\int_0^{\pi}\int_0^{2\pi}\omega(\theta,\phi_1,\ldots,\phi_{n-2})\,d\theta\,d\phi_1\ldots d\phi_{n-2}}$$

where

$$\omega(\theta,\phi_1,\ldots,\phi_{n-2}) = \sin^{n-2}(\theta)\,\Pi_{k=1}^{n-3}\sin^{n-2-k}(\phi_k)$$

for $n \geq 3$ (as in [160, pages 227-228]), and

$$\frac{\delta(r,s,\theta)}{\delta(r,q,\theta)} = \frac{\sqrt{s^2 - r^2\sin^2\theta}}{\sqrt{q^2 - r^2\sin^2\theta}}.$$

After simplifying, we have

$$p(s;r,q) = \frac{\displaystyle\int_0^{\theta_0}\frac{\sqrt{s^2 - r^2\sin^2\theta}}{\sqrt{q^2 - r^2\sin^2\theta}}\sin^{n-2}(\theta)\,d\theta}{\int_0^{\pi/2}\sin^{n-2}(\theta)\,d\theta}$$

and because, for $w, z > 0$ the Beta function can be written as [5, page 258],

$$\frac{1}{2}B(z,w) = \frac{1}{2}\frac{\Gamma(z)\Gamma(w)}{\Gamma(z+w)} = \int_0^{\pi/2}\sin^{2z-1}(t)\cos^{2w-1}(t)\,dt$$

we can write

$$p(s;r,q) = \frac{2}{B\left(\frac{n-1}{2},\frac{1}{2}\right)}\int_0^{\theta_0}\frac{\sqrt{\frac{s^2}{r^2} - \sin^2\theta}}{\sqrt{\frac{q^2}{r^2} - \sin^2\theta}}\sin^{n-2}(\theta)\,d\theta.$$

It is easily shown that this expression is also valid for $n = 2$ and 3.

We now concentrate on the integral and go through a series of substitutions in order to get the final expression. First, let $a = s^2/r^2$ and $b = q^2/r^2$ to get

$$\int_0^{\theta_0}\frac{\sqrt{\frac{s^2}{r^2} - \sin^2\theta}}{\sqrt{\frac{q^2}{r^2} - \sin^2\theta}}\sin^{n-2}(\theta)\,d\theta \quad = \quad \int_0^{\theta_0}\left(\frac{a - \sin^2\theta}{b - \sin^2\theta}\right)^{1/2}\sin^{n-2}(\theta)\,d\theta$$

and then change variables using $\sin^2\theta = at$ to get

$$= \int_0^1\left(\frac{a - at}{b - at}\right)^{1/2}\frac{(at)^{\frac{n-2}{2}}\,a\,dt}{2\,(at)^{1/2}\,(1 - at)^{1/2}}$$

$$= \frac{a^{\frac{n}{2}}}{2} \int_0^1 \left(\frac{1-t}{(b-at)(1-at)} \right)^{1/2} t^{\frac{n-3}{2}} \, dt$$
$$= \frac{a^{\frac{n-2}{2}}}{2} \int_0^1 \left(\frac{1-t}{(b/a-t)(1/a-t)} \right)^{1/2} t^{\frac{n-3}{2}} \, dt$$

which, using [57, 3.211 of page 287], reduces to

$$= \frac{a^{\frac{n}{2}}}{2 b^{\frac{1}{2}}} B\left(\frac{3}{2}, \frac{n-1}{2}\right) F_1\left(\frac{n-1}{2}, \frac{1}{2}, \frac{1}{2}, \frac{n+2}{2}; \frac{a}{b}, a\right).$$

Noticing that

$$\frac{B\left(\frac{3}{2}, \frac{n-1}{2}\right)}{B\left(\frac{n-1}{2}, \frac{1}{2}\right)} = \frac{1}{n}$$

we can write

$$p(s;r,q) = \frac{a^{\frac{n}{2}}}{b^{\frac{1}{2}}} \left(\frac{1}{n}\right) F_1\left(\frac{n-1}{2}, \frac{1}{2}, \frac{1}{2}, \frac{n+2}{2}; \frac{a}{b}, a\right)$$

and returning to the original variables

$$= \left(\frac{r}{q}\right) \left(\frac{s}{r}\right)^n \frac{1}{n} F_1\left(\frac{n-1}{2}, \frac{1}{2}, \frac{1}{2}, \frac{n+2}{2}; \frac{s^2}{q^2}, \frac{s^2}{r^2}\right). \quad \blacksquare$$

The special case where $s = r$, denoted $p(r;r,q)$, is the probability that a sample point is improving given the current point is at $Y_k = h(r)$. This case simplifies to (using [57, equation 10, page 1055]),

$$p(r;r,q) = \left(\frac{r}{q}\right) \frac{1}{\gamma(n)} F\left(\frac{1}{2}, \frac{n-1}{2}; \frac{n+1}{2}; \frac{r^2}{q^2}\right) \tag{6.5}$$

where

$$\gamma(n) = \frac{\Gamma(\frac{n+1}{2})\Gamma(\frac{1}{2})}{\Gamma(\frac{n}{2})}$$

for $0 < r \le q$, and for $k = 0, 1, 2, \ldots$, and $F(a, b; c; z)$ is Gauss' hypergeometric function (see [5] or [57]).

To measure complexity of IHR on a spherical program, define $N(r)$ as the number of sample points needed to achieve an objective function value less than or equal to $h(r)$ with $0 < r \le q$. The sample point count $N(r)$ is also equal to the number of function evaluations, and because each iteration has little computation other than the function evaluation, this is an accurate indication of total computing time. In the next result we also define an improving point count $K(r)$ as the number of improving points (records) needed to get below $h(r)$. The following theorem, from [189, Theorem 3.3] is an intermediate result that expresses the expected number of sample points in terms of the expected number of improving points.

THEOREM 6.6 *For any spherical program (SP), the expected number of IHR sample points needed to achieve an objective function value of $h(r)$ or better is bounded by*

$$\begin{aligned} E[N(r)] &\leq \frac{E[K(r)]}{p(r;r,q)} \\ &\leq \frac{q}{r}\gamma(n)\,E[K(r)] \end{aligned}$$

for $0 < r \leq q$ and where

$$\begin{aligned} \gamma(n) &= \frac{\Gamma(\frac{n+1}{2})\Gamma(\frac{1}{2})}{\Gamma(\frac{n}{2})} \\ &= O(\sqrt{n}). \end{aligned}$$

Proof: From [189], we first express $N(r)$ as a sum,

$$N(r) = \sum_{m=1}^{K(r)} N_m$$

where N_m is defined to be the number of sample points between consecutive improving points (including the sample point corresponding to the mth record). It is convenient to let $\{R_k\}$, $k = 0, 1, \ldots$ be the sequence of radii in a spherical program associated with the record values generated by IHR, namely,

$$R_k = ||X_{R(k)} - x_*||_I$$

where $X_{R(k)}$ is the kth record point. Also,

$$R_k = h^{-1}(Y_{R(k)}) = h^{-1}(f(X_{R(k)})).$$

Now we consider the expected value of $N(r)$ and proceed to condition on the record values to get,

$$\begin{aligned} E[N(r)] &= E\left[\sum_{m=1}^{K(r)} N_m\right] \\ &= E\left[E\left[\sum_{m=1}^{K(r)} N_m | K(r), R_0, \ldots, R_{K(r)-1}\right]\right]. \end{aligned}$$

The conditional expectation can be rewritten, for $r_0 > r_1 > \cdots r_{k-1} > r$, and $k \geq 1$:

$$
\begin{aligned}
& E\left[\textstyle\sum_{m=1}^{K(r)} N_m | K(r) = k, R_0 = r_0, \ldots, R_{k-1} = r_{k-1}\right] \\
&= E\left[\textstyle\sum_{m=1}^{k} N_m | K(r) = k, R_0 = r_0, \ldots, R_{k-1} = r_{k-1}\right] \\
&= E\left[\textstyle\sum_{m=1}^{k} N_m | R_0 = r_0, \ldots, R_{k-1} = r_{k-1}, R_k \leq r\right]
\end{aligned}
$$

because the kth improving point is the first point inside the r-sphere

$$= \sum_{m=1}^{k} E\left[N_m | R_0 = r_0, \ldots, R_{k-1} = r_{k-1}, R_k \leq r\right]$$

by linearity of conditional expectation

$$= \sum_{m=1}^{k} E\left[N_m | R_{m-1} = r_{m-1}\right]$$

because N_m is conditionally independent of R_j $(j \neq m-1)$ when given R_{m-1}

$$= \sum_{m=1}^{k} \frac{1}{p(r_{m-1}; r_{m-1}, q)}$$

because the number of sample points to achieve the mth improving point, given $R_{m-1} = r_{m-1}$, has a geometric distribution with parameter $p(r_{m-1}; r_{m-1}, q)$

$$\leq \sum_{m=1}^{k} \frac{1}{p(r; r, q)}$$

since $p(r; r, q) \leq p(s; s, q)$ for $s \geq r$

$$= \frac{k}{p(r; r, q)}.$$

We now can show the first statement in the theorem,

$$
\begin{aligned}
E[N(r)] &\leq E\left[\tfrac{K(r)}{p(r;r,q)}\right] \\
&= \frac{E[K(r)]}{p(r; r, q)}.
\end{aligned}
$$

The second statement follows easily from the expression for $p(r; r, q)$ as given in Equation 6.5,

$$E[N(r)] \leq \frac{q}{r}\gamma(n) F\left(\frac{1}{2}, \frac{n-1}{2}; \frac{n+1}{2}; \frac{r^2}{q^2}\right) E[K(r)]$$

where $\gamma(n) = \frac{\Gamma(\frac{n+1}{2})\Gamma(\frac{1}{2})}{\Gamma(\frac{n}{2})}$. Noting that $F\left(\frac{1}{2}, \frac{n-1}{2}; \frac{n+1}{2}; \frac{r^2}{q^2}\right) \leq 1$, we have

$$E[N(r)] \leq \frac{q}{r}\gamma(n) E[K(r)].$$

Now for large n, we have

$$\frac{\Gamma(\frac{n+1}{2})}{\Gamma(\frac{n}{2})} \approx \sqrt{n/2}$$

(see [38]), yielding $\gamma(n) = O(\sqrt{n})$. ∎

We now analyze the performance of improving points of IHR. We start by evaluating a conditional probability for improving points and then develop a bound on the expected number of improving points $E[K(r)]$ in terms of dimension, as in [189, Lemma 3.4].

LEMMA 6.7 *For any spherical program (SP), the conditional probability distribution of improving points for IHR is given by*

$$\begin{aligned} P(R_{k+1} < s | R_k = r, R_0 = q) &= \frac{p(s; r, q)}{p(r; r, q)} \\ &\geq \frac{(s/r)^n}{n} \end{aligned}$$

for $0 < s \leq r \leq q$, *and for* $k = 0, 1, 2, \ldots$.

Proof: This proof follows [189]. Suppose the kth record occurs at iteration j, then,

$$\begin{aligned} &P(R_{k+1} < s | R_k = r, R_0 = q) \\ &= P(Y_{j+1} < h(s) | Y_j = h(r), Y_0 = h(q)) \\ &\quad + P(Y_{j+2} < h(s), Y_{j+1} \geq h(r) | Y_j = h(r), Y_0 = h(q)) \\ &\quad + \cdots \\ &= p(s; r, q) + p(s; r, q)(1 - p(r; r, q)) + \cdots \\ &= p(s; r, q) \sum_{i=0}^{\infty} (1 - p(r; r, q))^i \\ &= \frac{p(s; r, q)}{p(r; r, q)}. \end{aligned}$$

From Lemma 6.5 and Equation 6.5 we have

$$p(s; r, q) = \left(\frac{r}{q}\right)\left(\frac{s}{r}\right)^n \frac{1}{n} F_1\left(\frac{n-1}{2}, \frac{1}{2}, \frac{1}{2}, \frac{n+2}{2}; \frac{s^2}{q^2}, \frac{s^2}{r^2}\right)$$

and

$$p(r; r, q) = \left(\frac{r}{q}\right) \frac{\Gamma(\frac{n}{2})}{\Gamma(\frac{n+1}{2})\Gamma(\frac{1}{2})} F\left(\frac{1}{2}, \frac{n-1}{2}; \frac{n+1}{2}; \frac{r^2}{q^2}\right).$$

It is easy to see from definitions that $F(a, b; c; z)$ is increasing in z for $z > 0$ when $a, b, c > 0$. Also, $F_1(a, b, c, d; x, y)$ is increasing in both x and y for $x, y > 0$ when $a, b, c, d > 0$. Thus,

$$\begin{aligned} F\left(\frac{1}{2}, \frac{n-1}{2}; \frac{n+1}{2}; \frac{r^2}{q^2}\right) &\leq F\left(\frac{1}{2}, \frac{n-1}{2}; \frac{n+1}{2}; 1\right) \\ &= \frac{\Gamma(\frac{n+1}{2})\Gamma(\frac{1}{2})}{\Gamma(\frac{n}{2})} \end{aligned}$$

from [57, 9.122.1 on page 1042] for $0 < r/q \leq 1$. Similarly,

$$\begin{aligned} F_1\left(\frac{n-1}{2}, \frac{1}{2}, \frac{1}{2}, \frac{n+2}{2}; \frac{s^2}{q^2}, \frac{s^2}{r^2}\right) &\geq F_1\left(\frac{n-1}{2}, \frac{1}{2}, \frac{1}{2}, \frac{n+2}{2}; 0, 0\right) \\ &= 1. \end{aligned}$$

Hence

$$\begin{aligned} \frac{p(s; r, q)}{p(r; r, q)} &= \left(\frac{s}{r}\right)^n \frac{1}{n} \frac{F_1\left(\frac{n-1}{2}, \frac{1}{2}, \frac{1}{2}, \frac{n+2}{2}; \frac{s^2}{q^2}, \frac{s^2}{r^2}\right)}{F\left(\frac{1}{2}, \frac{n-1}{2}; \frac{n+1}{2}; \frac{r^2}{q^2}\right)} \frac{\Gamma(\frac{n+1}{2})\Gamma(\frac{1}{2})}{\Gamma(\frac{n}{2})} \\ &\geq \frac{(s/r)^n}{n}. \end{aligned}$$

■

The following theorem shows that the expected number of improving points of IHR has a quadratic bound. This intermediate result is interesting compared to the linear bound of pure adaptive search. It says that even though IHR does not sample according to a true uniform distribution on improving sets, the degradation in expected number of improving points is not severe. The following theorem first appeared in [189, Theorem 3.5].

THEOREM 6.8 *For any spherical program (SP), the expected number of IHR improving points needed to achieve an objective function value of*

$h(r)$ or better is bounded by

$$\begin{aligned} E[K(r)] &\leq nE\left[K(r)^{PAS}\right] \\ &= O(n^2) \end{aligned}$$

for $0 < r \leq q$ where $K(r)^{PAS}$ is the corresponding expected number of improving points for PAS.

Proof: From Lemma 6.7 we have a bound on the conditional probability for improving points for IHR, for $0 < s \leq r \leq q$

$$P\left(R_{k+1}^{IHR} < s | R_k^{IHR} = r\right) \geq \frac{(s/r)^n}{n}.$$

Similarly, for pure adaptive search (PAS) we have [188],

$$P\left(R_{k+1}^{PAS} < s | R_k^{PAS} = r\right) = (s/r)^n.$$

Now we define the following intermediate algorithm, called A.

Algorithm A

Step 0. Initialize $X_0 \in S$, and $k = 0$.

Step 1. With probability $1/n$, choose X_{k+1} uniformly from the set

$$\{x : x \in S, \text{and } f(x) < f(X_k)\}.$$

Otherwise, set $X_{k+1} = X_k$.

Step 2. Increment k, and return to Step 1.

Algorithm A performs a PAS step with probability $1/n$. Thus for $0 < s \leq r \leq q$ we have,

$$P\left(R_{k+1}^{A} < s | R_k^{A} = r\right) = \frac{(s/r)^n}{n}$$

and further,

$$P\left(R_{k+1}^{A} < s | R_k^{A} = r\right) \leq P\left(R_{k+1}^{IHR} < s | R_k^{IHR} = r\right).$$

We have defined $E[K(r)]$ to be the expected number of improving points needed to achieve $R_k \leq r$, and we now extend the definition for the three algorithms and add a superscript. We now have $E\left[K(r)^{IHR}\right]$,

$E\left[K(r)^{PAS}\right]$, and $E\left[K(r)^{A}\right]$. With the above comments, we have the following,

$$\begin{aligned} E\left[K(r)^{IHR}\right] &\leq E\left[K(r)^{A}\right] \\ &= nE\left[K(r)^{PAS}\right] \\ &= O(n^2). \end{aligned}$$

■

The culminating result of polynomial complexity for IHR follows from combining Theorems 6.6 and 6.8 with Theorem 6.4. It first appeared in [189, Corollary 3.6].

Corollary 6.9 *For any elliptical program (EP), the expected number of IHR sample points needed to achieve an objective function value of $h(r)$ or better is bounded by*

$$\begin{aligned} E[N(r)] &\leq \frac{q}{r}\gamma(n)\, n\, E\left[K(r)^{PAS}\right] \\ &= O(n^{5/2}) \end{aligned}$$

for $0 < r \leq q$.

Proof: From Theorem 6.6 we have, for $0 < r \leq q$,

$$E[N(r)] \leq \frac{q}{r}\gamma(n)\, E[K(r)]$$

and from Theorem 6.8

$$\leq \frac{q}{r}\gamma(n)\, n\, E\left[K(r)^{PAS}\right]$$

which yields

$$= O(n^{5/2}).$$

■

Although the intent is not to apply IHR to quadratic programs, the polynomial complexity result appears to hold even if the conditions are not exactly satisfied. Suppose a global optimization problem has very small perturbations in the objective function with many local minima, but the overall function has a strong underlying convex shape. Then one might expect that the performance of IHR should be close to the above theoretical complexity.

2.3 Discussion

Improving Hit-and-Run has been shown to have a search effort that is polynomially bounded in dimension for the class of elliptical programs. This complexity is attainable for strictly convex quadratic programs by choosing H to be the Hessian of the objective function. Typically the Hessian is not known, but many approximating methods have been developed in the context of quasi-Newton local search algorithms (see e.g. [146]) which can be applied to IHR.

The motivation for Improving Hit-and-Run is that its computational complexity is comparable to pure adaptive search. If extremely long sequences of Hit-and-Run are used to approximate PAS, then we expect a linear bound in dimension on improving points at a comparably high cost of obtaining an improving point. For IHR, which employs only one step of Hit-and-Run per iteration, the expected number of improving points has a quadratic bound (Theorem 6.8) which although worse than PAS is associated with a comparably low cost of obtaining an improving point. In fact, the total expected number of function evaluations is only $O(n^{5/2})$ on elliptical programs. Computational experience with IHR indicates that its performance on nonelliptical programs is relatively insensitive to small perturbations in an objective function that has an underlying convex function. Numerical results are summarized in Section 5. The combination of computational and theoretical results on the performance of Improving Hit-and-Run provide support for the value of embedding Hit-and-Run within an optimization framework.

3. Hide-and-Seek

Hide-and-Seek, proposed by Romeijn and Smith [140], generalizes Improving Hit-and-Run by adding a Metropolis acceptance criterion. The motivation behind IHR is to emulate PAS, but in practice IHR tends to get trapped at local optima. One way to alleviate this is to accept non-improving points, consequently Hide-and-Seek. Hide-and-Seek is intended to emulate annealing adaptive search (see Chapter 4). Annealing adaptive search takes an alternative approach to PAS, and instead of generating uniform points only in the improving region it relaxes that goal and instead generates points from a nonuniform distribution over the entire feasible region. Annealing adaptive search generates points drawn from a sequence of Boltzmann distributions, parameterized by temperature T, on the full feasible region S. Romeijn and Smith [139] showed that the record values of annealing adaptive search stochastically dominate pure adaptive search, and thus the improving points inherit the linearity property. Hide-and-Seek [140] uses Hit-and-Run with a Metropolis acceptance/rejection criterion to approximate the Boltzmann distributions of annealing adaptive search. Hide-and-Seek has been shown to converge to the global optimum almost surely under certain conditions on the cooling schedule (see Section 3.3).

3.1 Definition of Hide-and-Seek

Hide-and-Seek is a simulated annealing algorithm where the generator of the algorithm is Hit-and-Run. One of the advantages of using Hit-and-Run as a generator within simulated annealing is that it involves no parameters that need fine tuning; all it needs is a random direction and a random step length. In this sense it is a robust generator. While most simulated annealing generators are local in the sense that only immediate neighbors or a small subset of the domain has a possibility of being generated, Hit-and-Run has a positive probability of sampling anywhere in the domain on a single iteration.

Hide-and-Seek is defined on the same global optimization problem stated in Section 2.1,

$$(P) \qquad \min_{x \in S} f(x)$$

where f is a real-valued function on S, and the domain S is assumed to be a compact, full-dimensional subset of $\mathbb{R}^n$, whose boundary has Lebesgue measure zero.

We define the Hide-and-Seek algorithm based on [140].

Hide-and-Seek

Step 0. Initialize $X_0 \in S$, $Y_0 = f(X_0)$, and $T_0 \in (0, \infty)$. Set $k = 0$.

Step 1. Generate a random direction vector D_k according to a uniform distribution on a unit n-dimensional hypersphere.

Step 2. Generate a candidate point $W_{k+1} = X_k + \lambda D_k$ by sampling uniformly over the line set

$$L_k = \{x : x \in S \text{ and } x = X_k + \lambda D_k, \lambda \text{ a real scalar}\}.$$

If $L_k = \emptyset$, go to Step 1.

Step 3. Update the current point X_{k+1} by accepting or rejecting the candidate point as follows,

$$X_{k+1} = \begin{cases} W_{k+1} & \text{with probability } P_{T_k}(X_k, W_{k+1}) \\ X_k & \text{otherwise} \end{cases}$$

where $P_{T_k}(X_k, W_{k+1}) = \min\{1, \exp(f(X_k) - f(W_{k+1}))/T\}$.

Update the temperature according to a cooling schedule, i.e. set $T_{k+1} = \tau(T_k)$, and set $Y_{k+1} = f(X_{k+1})$. Also update the best point so far, $Y_{Best} = \min\{Y_{Best}, Y_{k+1}\}$.

Step 4. If the stopping criterion is met, stop. Otherwise increment k and return to Step 1.

The difference between Hide-and-Seek and Improving Hit-and-Run is in the acceptance criterion and cooling schedule. Otherwise the implementation issues are the same as IHR (see Section 2.1).

3.2 Acceptance Criterion and Cooling Schedule

In Step 3 of Hide-and-Seek, the probability of accepting the candidate point W, given the current iteration point X and the current temperature T, is expressed as $P_T(X, W)$. This probability of accepting non-improving points is characteristic of simulated annealing algorithms. Simulated annealing methods that generate only immediate neighbors as candidate points rely on accepting non-improving points to escape a basin surrounding a local minimum. Even though Hit-and-Run has a positive probability of sampling anywhere in the space on one iteration, an acceptance probability is still useful in altering the asymptotic distribution of the points generated.

For Hide-and-Seek, the acceptance probability is

$$P_T(X, W) = \begin{cases} 1 & \text{if improving, } f(W) \le f(X), \\ \exp\left\{\frac{f(X) - f(W)}{T}\right\} & \text{otherwise} \end{cases}$$

also known as the Metropolis criterion [110]. For a constant temperature T, the iteration points converge to a Boltzmann distribution with density

$$g_T(x) = \frac{e^{-f(x)/T}}{\int_S e^{-f(z)/T} dz}.$$

Thus a sequence of Hide-and-Seek at a fixed temperature is a means of approximating a Boltzmann distribution. As the temperature follows the cooling schedule, Hide-and-Seek can be viewed as approximating a sequence of Boltzmann distributions, parametrized by temperature T, and is an approximation of annealing adaptive search. If it is a good approximation of annealing adaptive search, then we expect it to have nearly the same promising performance.

An interpretation of the role of the temperature parameter in the Boltzmann distribution is to focus the sampling distribution closely around the global optimum. When the temperature is high, most of the random moves in the domain are accepted. In the extreme, when the acceptance probability always equals one, loosely when $T = \infty$, then all candidate points are accepted. In this situation, the Hit-and-Run generator converges to a uniform distribution, and keeping $T = \infty$ reduces Hide-and-Seek to pure random search. At the other extreme, when only improving points are accepted, loosely when $T = 0$, then Hide-and-Seek becomes IHR. This provides another interpretation of IHR as an attempt to approximate the degenerate Boltzmann distribution with all the probability mass at the global optimum. The idea is that the computation involved in approximating a series of Boltzmann distributions is an effective approximation of annealing adaptive search.

The question remains of how to gradually reduce the temperature T in a cooling schedule. A common cooling schedule used in simulated annealing is a simple geometric function, $T_{k+1} = C \cdot T_k$ where C is a constant between 0 and 1. The temperature may be adjusted every N_T iterations, rather than every iteration. The specific constants for the cooling schedule are often chosen based on experience with a particular problem.

The cooling schedule in Hide-and-Seek is theoretically motivated by the analysis of annealing adaptive search. Cooling schedules for annealing adaptive search are discussed in Chapter 4, Section 3. Romeijn and Smith [140] showed that virtually any cooling schedule for Hide-and-Seek can be chosen, as long as the temperature converges to zero in probability. Therefore it is possible to adapt a cooling schedule derived for annealing adaptive search to be used in Hide-and-Seek, and use estimates that are based on the progress of the algorithm. Romeijn and Smith [140] conducted numerical experiments with five cooling schedules of Hide-and-Seek. These computational results are summarized in Section 5.

3.3 Convergence of Hide-and-Seek

Hide-and-Seek has been shown to converge in probability to the global optimum regardless of the rate of cooling. This may be surprising when compared to similar results for simulated annealing in the discrete case which specify conditions on the cooling schedule to get convergence results [1]. However, an important difference is that the Hit-and-Run generator has a positive probability of reaching anywhere in S on a single iteration, in contrast to a restricted local neighborhood search.

The following theorem is due to Bélisle [14].

THEOREM 6.10 *Consider the sequence $\{X_k, k = 0, 1, \ldots\}$ generated by the Hide-and-Seek algorithm using a cooling schedule $\{T_k, k = 0, 1, \ldots\}$. Assume that for every starting point x_0, the cooling schedule converges to 0 in probability. Then the sequence of objective function values $\{f(X_k), k = 0, 1, \ldots\}$ converges in probability to the global minimum y_*. That is, for all $\epsilon > 0$ and $x_0 \in S$,*

$$P(f(X_k) > y_* + \epsilon | X_0 = x_0) \to 0 \quad as \quad k \to \infty.$$

Proof: See [14] for the proof of a slightly more general result with several convergence conditions that Hide-and-Seek satisfies. ∎

A slightly weaker form of convergence is to investigate the best point sampled, in contrast to the last point sampled in the sequence. For instance, if Hide-and-Seek uses a constant temperature value that is very large, it behaves like pure random search, which converges to the global optimum when keeping track of the best point sampled. The best objective function value sampled by Hide-and-Seek converges to the global minimum [140],

$$Y_{Best} \to y_* \text{ almost surely, as } k \to \infty$$

for a broad class of problems, cooling schedules and direction distributions.

It is interesting that the convergence result does not put strong restrictions on how quickly the cooling schedule converges to zero. It is important to realize that the cooling schedule has an effect on the *rate* of convergence of Hide-and-Seek. The convergence rate also depends on, of course, the specific objective function and constraints in the problem. Analytical results for the performance of Improving Hit-and-Run (where the temperature is fixed at zero) on a special class of problems are summarized in Section 2.2. A summary of computational results for IHR and Hide-and-Seek is presented in Section 5.

4. Extensions to Hit-and-Run Based Optimization Methods

Extensions to Hit-and-Run, Improving Hit-and-Run and Hide-and-Seek have been developed in a variety of ways to improve performance. The following section describes alternative direction distributions that have been explored. Another extension has been to include optimization problems with both continuous and discrete variables, discussed in Section 4.2. It is not always possible to analyze the performance of these extensions analytically, and computational results are presented in Section 5.

4.1 Variations to Direction Generator

There are several possible direction sets, and consequently direction distributions, that may be used in Step 2 of Hit-and-Run, Improving Hit-and-Run and Hide-and-Seek. The hyperspherical direction (HD) set is straightforward to implement and has been widely used. For applying Improving Hit-and-Run to elliptical programs, as seen in Section 2.1, the direction distribution may be generated from a multivariate normal distribution with mean zero and covariance matrix H^{-1}. The disadvantage is that the Hessian is not typically known.

Another natural choice for direction distribution is the coordinate direction (CD) method, where the direction is chosen uniformly from the n coordinate vectors (spanning $\mathbb{R}^n$). Both HD and CD versions of direction choice were presented and applied to identifying nonredundant linear constraints in Berbee et al. [19]. While Improving Hit-and-Run and Hide-and-Seek are guaranteed to converge with probability one for a wide class of problems when using the hyperspherical direction choice, the same is not true when using the coordinate direction choice. It is possible to construct problems where the algorithms with CD may not converge to the global optimum by shifting the alignment of local and global optima. Kristinsdottir provides a simple example in [93] where local minima are lined up on the coordinate directions, and it is impossible for Improving Hit-and-Run with the CD method to leave the local minimum when accepting only improving points. If the algorithm with CD is coupled with a positive acceptance probability for non-improving points, then it converges with probability one to the global minimum. Computational experiments exploring the tradeoff between HD and CD with various acceptance probabilities were presented in [93]. The numerical experiments were performed on a discretized quadratic function $\min \sum_{i=1}^{n} x_i^2$ over a discrete set within $\{-K, +K\}$, where $n = 2, 3, 10, 30,$and 50 and $K = 1, 2, 5, 25,$ and 50. For this test function, CD consistently outperformed HD, especially for the problems with larger values for n and K. It may happen that a problem is oriented along the coordinate directions in such a way that a coordinate direction choice is more likely to progress rapidly towards the global optimum.

Neogi [116] numerically tested a variation on coordinate direction choice, called cyclic-CD, where the coordinate directions are generated sequentially in their natural order, from 1 to n. Initial computational results indicated that cyclic-CD is more effective than HD, however, again, IHR with cyclic-CD can get trapped easily, while IHR with HD converges with probability one to the global minimum.

Kaufman and Smith [81] derived a non-uniform direction distribution that optimizes the rate of convergence of Hit-and-Run to a target distribution. They show that under regularity conditions on the region S and the target distribution π, there exists a unique direction distribution which optimizes a bound on the rate of convergence. To say it simply, the best choice of direction distribution is based on knowing the Hessian matrix at the optimal point. Although exact sampling from the optimal direction distribution may be very difficult, it motivates a heuristic adaptive direction choice rule called Artificial Centering Hit-and-Run. In Artificial Centering Hit-and-Run, directions are chosen by randomly selecting one of the the previous Hit-and-Run iterates and normalizing the vector from an approximate center to the iterate. Although the results in [81] are for Hit-and-Run, they can easily be used to provide a direction distribution in Hide-and-Seek. As summarized in Section 5, promising results were found in numerical experiments using this type of approximation in a direction choice.

Another modification of the direction choice, introduced by Romeijn et al. [141], is called a reflection generator. The reflection generator was motivated by the problem of stalling discussed in Section 1.1 and illustrated in Figure 6.1. An observation of Hit-and-Run on a hypercube is that when the current point x is near a corner, there is a high probability that the next sample point is very close to x, and a very low probability that the next point generated is a substantial distance from x. This problem is amplified when the number of dimensions is large. This problem is similar to jamming, a well-known problem in nonlinear programming, and has been discussed in the context of feasible directions methods (see [192]).

To partially alleviate the problem of jamming, the idea of reflection was developed. The reflection generator essentially lengthens the line associated with a chosen direction by reflecting it off the boundaries of the feasible region into the interior. This increases the probability of sampling a point far away from the current point. A general reflection generator is defined in [141]. Convergence results similar to Theorem 6.10 are proven for a variation to Hide-and-Seek that uses a general reflection generator.

A straightforward example of a reflection generator is component-by-component reflection. Suppose the feasible region is an n-dimensional box with lower bounds denoted x_i^L and upper bounds x_i^U for $i = 1, \ldots, n$. One way to implement component-by-component reflection is to embed the original box in a larger one, generate a random direction and point within the larger box, and then map the point back into the original box. For example, arbitrarily double the length of each side so that lower and upper bounds on the larger box are given by

$$z_i^L = x_i^L - (1/2)(x_i^U - x_i^L)$$

$$z_i^U = x_i^U + (1/2)(x_i^U - x_i^L)$$

for $i = 1, \ldots, n$. Suppose a random direction is chosen from a current point x in the original box and z is the random point generated. The new point z must be contained in the larger box, but may not be contained in the original box. The point z is mapped back into the original box using $m(z)$, so that

$$m(z_i) = \begin{cases} x_i^L + (x_i^L - z_i) & \text{if } z_i < x_i^L, \\ z_i & \text{if } x_i^L \leq z_i \leq x_i^U, \\ x_i^U - (z_i L - x_i^U) & \text{if } z_i > x_i^U \end{cases}$$

for $i = 1, \ldots, n$. Notice that $m(z)$ is contained in the original box. This component-by-component mapping is easy to implement and captures the geometric idea of reflection. Positive numerical experience was reported in [141] and is summarized in Section 5.

4.2 Discrete Variations of Hit-and-Run

Hit-and-Run was originally defined on a continuous set, and its incorporation into Improving Hit-and-Run and Hide-and-Seek has been in the context of a global optimization problem with continuous variables. However many practical problems involve both continuous and discrete variables, so several variations of Hit-and-Run have been developed to accommodate problems with both continuous and discrete variables.

The use of Hit-and-Run as a generator in an optimization algorithm was motivated by the linear performance of pure adaptive search in a continuous domain. Fortunately, there is an analogous linearity property of pure adaptive search in a discrete domain. This motivates a modification of Hit-and-Run created for a discrete or mixed continuous-discrete domain.

Three approaches of adapting the basic Hit-and-Run algorithm to discrete domains are presented in [141], [117] and [85, 87]. The first extension to a discrete domain is accomplished by superimposing the discrete domain onto a continuous domain [141], for example, placing the integer points in a hypercube in $\mathbb{R}^n$. This was motivated by engineering problems, where discrete design variables, such as fiber angles in a composite laminate, or diameters in a 10-bar truss, have a natural continuous analog. This variation is called the *step-function approach.* The second extension is similar to the step-function approach, but each iteration in the continuous domain is rounded to an integer point [117], hence called the *rounding approach.* The third version is a more direct approach of generating a discrete point [85, 87], and is called *discrete biwalk Hit-and-Run.*

Step-function Approach. The step-function approach introduced by Romeijn et al. [141] reformulates the original mixed continuous-integer global optimization problem into a purely continuous one. Then Improving Hit-and-Run or Hide-and-Seek is applied to the purely continuous problem, and the optimal solution is transformed back into the original variables. The Hit-and-Run generator is always generating candidate points on a continuous domain, while the objective function value of a non-integer point is set equal to the objective function evaluated at its nearest integer neighbor. In this way, the algorithm operates on a continuous domain and the objective function is a multi-dimensional step function, with plateaus surrounding the integer points. This has been straightforward to implement, and has been tested on discretized sinusoidal functions, as well as on an engineering application for the design of composite structures discussed in Chapter 7. In that application, the discrete variables are fiber angles which are restricted to a set of angles, but theoretically could range between 0 and 180 degrees, so the step-function approach is appropriate.

The reformulation of the mixed continuous-integer problem is done in two steps. First the domain is extended, and second the step function is created. Suppose the original problem is

$$(P) \qquad \begin{array}{ll} \text{minimize} & f(x,y) \\ \text{subject to} & (x,y) \in S \end{array}$$

where x is an n-dimensional real vector, y is an m-dimensional integer vector, $f(x,y)$ is a real-valued function defined on S, and

$$S \subseteq \{(x,y) : x \in \mathbb{R}^n, y \in \mathbb{Z}^m, l \leq x \leq u, L \leq y \leq U\}$$

with $l < u \in \mathbb{R}$ and $L \leq U \in \mathbb{Z}$.

The reformulated problem has all continuous variables, and is

$$(P') \qquad \begin{array}{ll} \text{minimize} & \hat{f}(x,y) \\ \text{subject to} & (x,y) \in \hat{S} \end{array}$$

where

$$\hat{S} = \text{closure}(\{(x,y) : x \in \mathbb{R}^n, y \in \mathbb{R}^m, (x, \lfloor y \rfloor) \in S\})$$

and

$$\hat{f}(x,y) = f(x, \min\{U, \lfloor y \rfloor\}).$$

This reformulation effectively replaces every $(x,y) \in S$ with a box and the reformulated objective function $\hat{f}$ is constant over the box.

This approach of creating plateaus, or a step function, and overlaying a continuous domain onto a discrete one, implies that the objective function is no longer continuous. However, the Hide-and-Seek algorithm still converges in probability to the global optimum on the reformulated problem, and the transformation back provides the global optimum to the original problem. Sufficient conditions for convergence of an algorithm of this form are given in [141], and rely on results in [14]. The algorithm using the reflection generator is also shown to converge with probability one to the global optimum. Computational results reported in [141] indicate that the reflection generator on a reformulated problem outperforms that without using reflection. It has also been widely used in applications to composite design.

Rounding Approach. A second scheme to discretize Hit-and-Run suggested by Neogi [116, 117] is to similarly generate a point on a continuous domain, and then round the generated point to its nearest discrete point in the domain on each iteration. In this variation only discrete points are maintained. Although this approach has not been shown to converge in general, Neogi [116, 117] provided numerical results that indicate this approach is interesting. One advantage to the practitioner is that the current point is always a valid discrete point.

The diagram in Figure 6.6 illustrates a one-step transition for the the step-function approach and the rounding method. Starting from point X_1, Hit-and-Run on the purely continuous domain generates a candidate point such as A. The objective function at A is set equal to that of its nearest discrete point B, forcing $f(A) = f(B)$. If the candidate point is accepted, then $X_2 = A$, and another candidate point (shown as C) is generated.

If the rounding scheme is used, then the second point X_2 may be different. Again starting from point X_1 in Figure 6.6 suppose A is generated as a candidate point. The rounding scheme would replace A with its nearest discrete neighbor, in this example B. The objective function is evaluated at B, $f(B)$, and if the point is accepted, then $X_2 = B$. The difference at this point is small because $f(A) = f(B)$. However in the next iteration, a candidate point is generated from B instead of from A, see point D in the figure. The one-step transition probabilities for the two approaches are different. It is not clear which method is superior.

Discrete Biwalk Hit-and-Run. Research is ongoing [87, 85] to define a version of discrete Hit-and-Run with analytical properties analogous to the continuous form of Hit-and-Run. Preliminary results include definition of a biwalk on a discrete domain that can be shown to converge to a target distribution.

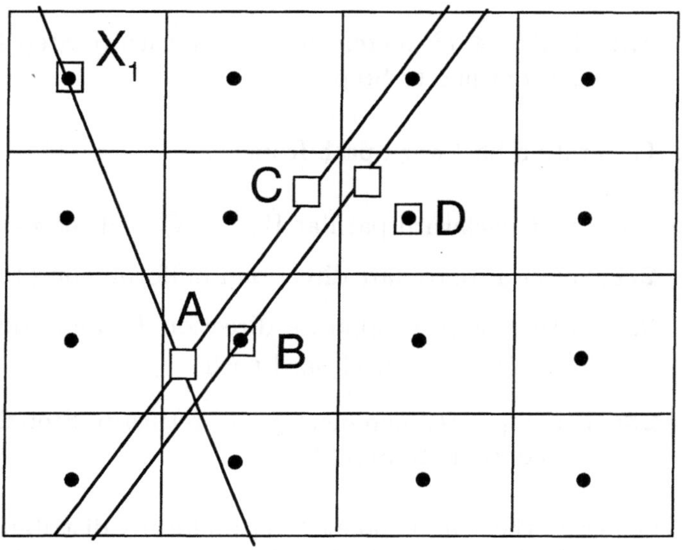

Figure 6.6. Starting at X_1, the step-function approach produces points A and C, while the rounding approach produces points B and D.

The optimization algorithm has the same form of Improving Hit-and-Run and Hide-and-Seek, whereas now a candidate point is generated using discrete biwalk Hit-and-Run. Once the candidate point is generated, the objective function is evaluated and the point may be accepted or rejected. So it remains to specify the generator.

Suppose we have a discrete set S that is a subset of $\mathbb{Z}^n$ and is contained in a hyperrectangle $H = \{(x_1, x_2, \ldots, x_n) \in \mathbb{Z}^n : 0 \leq x_i \leq l_i\}$. The discrete biwalk Hit-and-Run generates two random walks on the discrete domain, until they leave the hyperrectangle. Then the next point is sampled uniformly from the points along the forward path and the backward path. The discrete biwalk Hit-and-Run generator based on [87] follows.

Discrete Biwalk Hit-and-Run

Step 0. Initialize $X_0 \in S \subseteq H$. Set D_i equal to $+1$ or -1 with equal probability for each dimension i, $i = 1, \ldots, n$. Set $k = 0$.

Step 1. Generate a forward path from X_k.

1.0 Start forward path at $W_0^F = X_k$ and set $j = 0$.

1.1 Select a coordinate direction i uniformly on $\{1, 2, \ldots, n\}$.

1.2 Take a step in direction D_i and add $W_{j+1}^F = W_j^F + D_i$ to the forward path.

1.3 If $W^F_{j+1} \in H$, increment j and return to Step 1.1, otherwise continue to Step 2.

Step 2. Generate a backward path from X_k.

2.0 Start backward path at $W^B_0 = X_k$ and set $j = 0$.

2.1 Select a coordinate direction i uniformly on $\{1, 2, \ldots, n\}$.

2.2 Take a step in opposite direction D_i and add $W^B_{j+1} = W^B_j - D_i$ to the backward path.

2.3 If $W^B_{j+1} \in H$, increment j and return to Step 2.1, otherwise continue to Step 3.

Step 3. Generate the sample point X_{k+1} uniformly distributed on the set of distinct points included in the forward and backward paths $\left\{W^F_1, W^F_2, \ldots, W^B_1, W^B_2, \ldots\right\}$ that belong to S.

Step 4. If applying the Metropolis filter, update the current point X_{k+1} with the sample point with probability $\beta(X_k, X_{k+1}) = \min\{1, g(X_{k+1})/g(X_k)\}$, otherwise reject the sample point and set $X_{k+1} = X_k$.

Step 5. If a stopping criterion is met, stop. Otherwise increment k and return to Step 1.

If not applying the Metropolis filter in Step 4, or equivalently if $\beta = 1$, the Markov chain of points generated by discrete biwalk Hit-and-Run converges to a uniform distribution. If the Metropolis filter is used, the points converge to the target distribution defined by the density $g(x)$ used in the filter. The proofs are in [87] and rely on showing that discrete biwalk Hit-and-Run satisfies two properties; the transition density is symmetric, and has a positive probability of sampling anywhere in S. These results are analogous to the convergence properties of Hit-and-Run on a continuous domain.

The rate of convergence to the target distribution is also of interest. For the special case of approximating a uniform distribution on a bounded lattice of dimension n, Corollary 3 in [87] gives a bound of polynomial order n^5 for the rate of convergence in distribution. This version of discrete Hit-and-Run and slight variations have been embedded into a discrete form of Hide-and-Seek and computational results have been encouraging [155].

5. Computational Results

Computational results on Improving Hit-and-Run and Hide-and-Seek are reported in several papers, and are summarized here. It is difficult to make strong conclusions from numerical results. One caution is that the conclusions are specific to the test problems used, the number of repetitions chosen, and other details of how the experiments are conducted. Nonetheless, computational results complement theoretical results because it is possible to explore areas where the theoretical assumptions are not satisfied. It is also possible to investigate variations of algorithms where no theoretical analysis exists.

A discussion regarding comparison of global optimization software appears in Khompatraporn, et al. [84] and in Pintér [128]. The authors suggest that several key factors must be considered in evaluating the quality of global optimization software, including; generality, reliability, efficiency, and ease of use. In this section, we restrict ourselves to modifications of similar algorithms, and only consider reliability and efficiency. It is also suggested that numerical experiments have a specific objective or hypothesis to test. One theme here is to test how well Hit-and-Run based optimization algorithms approximate pure adaptive search and annealing adaptive search.

A set of experiments reported in [185] compared Improving Hit-and-Run with two direction generators, HD and CD, on a sinusoidal test function. The tests were run for problems of $n = 5, 10, 15, 20$, and 25 dimensions to compare with the linearity result of pure adaptive search.

The sinusoidal function was created with two purposes in mind. One is that it can be easily generalized to n dimensions, and has several parameters that can be controlled to shift the local of the global optimum, the relative depth of the local optima, and the frequency of the local optima. An expression for the number of local optima is given in [185] which grows rapidly (roughly exponentially) in dimension. The second purpose is that the sinusoidal function has a similarity to a realistic function important to the design of composite laminates. The optimization of composite structures is described in Chapter 7.

The sinusoidal function is stated as:

$$\text{maximize} \quad f(x) = A\prod_{i=1}^{n} \sin(x_i - C) + \prod_{i=1}^{n} \sin\left(B(x_i - C)\right)$$

$$\text{subject to} \quad 0^\circ \le x_i \le 180^\circ \quad \text{for} \quad i = 1, \ldots, n$$

where the parameter A affects the amplitude of the global optimum; B affects the periodicity and hence the number of local optima; C shifts the location of the global optima; and n indicates the dimension. The location of the global solution is at $x^* = (90+C, 90+C, \ldots, 90+C)$ with the global optimal value of $f(x^*) = A+1$. The number of local optima, for $B=5$ as given in [185], is

$$\sum_{i=0}^{\lfloor n/2 \rfloor} \left(\frac{n!}{(n-2i)!(2i)!} 3^{n-2i} 2^{2i} \right).$$

Figure 6.7 illustrates the sinusoidal function in two dimensions. The top graph has the optimum in the center because $C = 0$ degrees, while the bottom graph is shifted with $C = 30$ degrees. Both have $A = 2.5$ and $B = 5$.

For the experiments reported in [185], the record values were plotted to compare to the linearity prediction of pure adaptive search. IHR with both hyperspherical directions and coordinate directions was run on dimensions $n = 5, 10, 15, 20$, and 25. The number of improving points for IHR with both HD and CD appears to be linear in dimension. The number of function evaluations were also plotted, and although the sinusoidal function is not an elliptical program, the number of function evaluations for the results reported could be fit to a quadratic function. It was noted that as the relative depth of the local minima go to zero (with parameter A), the sinusoidal function approaches a spherical program. An interesting result is that the number of record values for IHR with CD is consistently less than with HD, however the number of function evaluations with HD is consistently less than with CD. Roughly, HD is more robust than CD with a higher success rate over the test problems. Although CD often gets stuck in a local minimum, it is interesting that when it is successful it uses less computation than HD.

A study with Hide-and-Seek was conducted and reported in [140]. Hide-and-Seek was run on nineteen test problems with three different direction distributions and five different cooling schedules. The test problems included box constraints with $l_i \leq x_i \leq u_i$ for $i = 1, \ldots, n$. Averages over 20 runs of each test problem were reported.

The three direction distributions are:

D1. Hyperspherical direction (HD). This generates a direction uniform on the boundary of the unit hypersphere and is equivalent to $D \sim N(0, I)$.

D2. Scaled direction. This generates a direction uniform on the boundary of the unit hypersphere after scaling the variables using their upper and lower bounds, and is equivalent to $D \sim N(0, B)$, where $B = \text{diag}((u_1 - l_1)^2, \ldots, (u_n - l_n)^2)$.

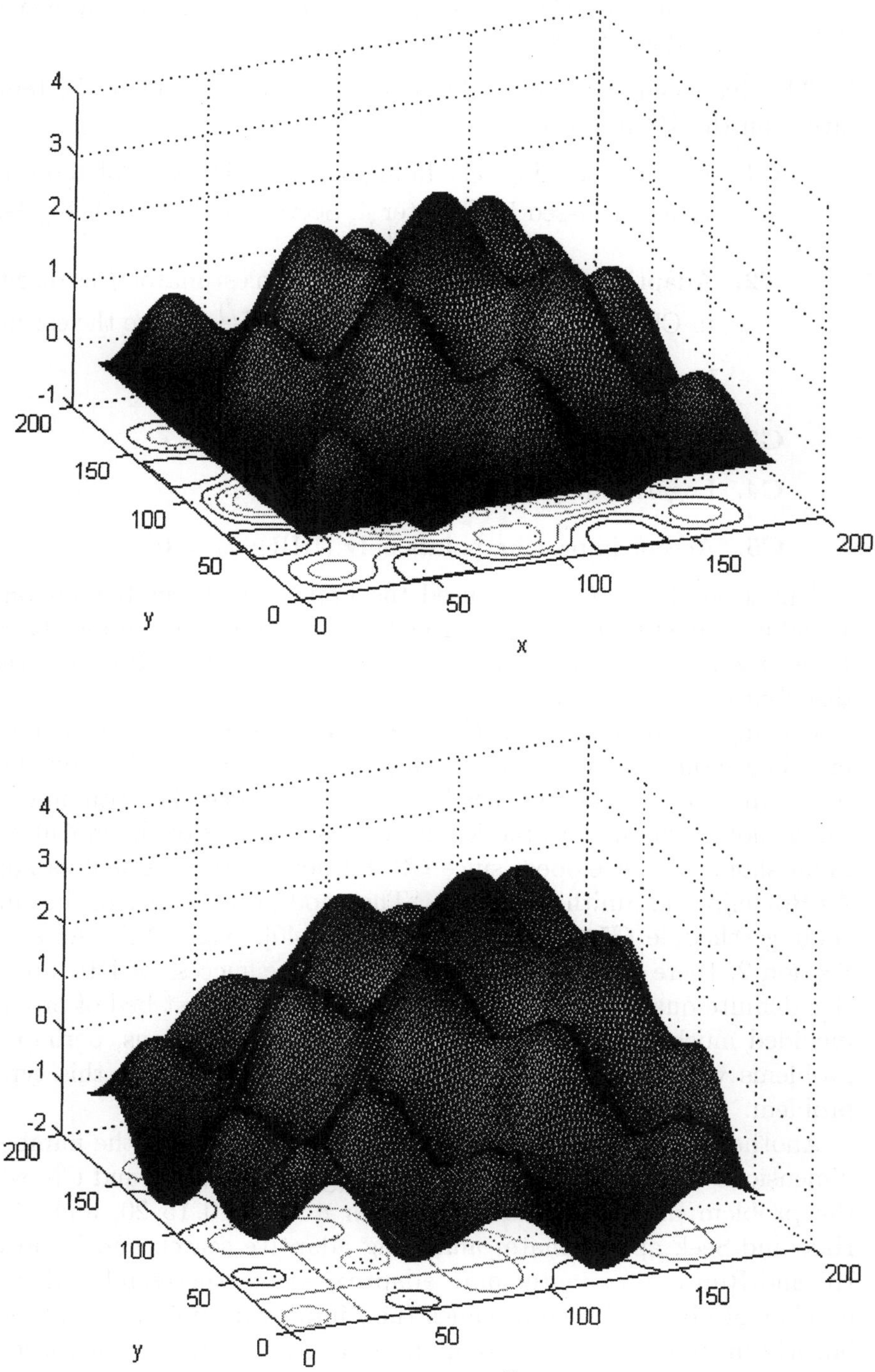

Figure 6.7. The sinusoidal function with $n = 2$, $A = 2.5$, $B = 5$, is shown centered ($C = 0°$) in the top graph, and shifted ($C = 30°$) in the bottom graph.

D3. D_i uniform in $(l_i - u_i, u_i - l_i)$ for $i = 1, \ldots, n$. This is motivated by the direction distribution developed by Kaufman and Smith [81] and summarized in Section 4.1.

The five cooling schedules that determine how to update the temperature on the kth iteration are:

C1. Adaptive cooling scheduling using the known global optimum f_* as discussed in Chapter 4, Section 3; $T_k = \frac{2(f(X_k)-f_*)}{\chi^2_{1-\alpha}(n)}$.

C2. Adaptive cooling scheduling using an estimator $\hat{f}$ as discussed in Chapter 4, Section 3; $T_k = \frac{2(f(X_k)-\hat{f})}{\chi^2_{1-\alpha}(n)}$, where the estimator is $\hat{f}(X_0, \ldots, X_k) = Y_{(k)} - \frac{Y_{(k-1)}-Y_{(k)}}{(1-\alpha)^{-n/2}-1}$.

C3. Geometric; $T_k = 0.99^k$.

C4. Logarithmic; $T_k = \frac{1}{\ln(k+1)}$.

C5. Accept improving points only (IHR); $T_k = 0$.

Direction choice $D3$, that used the upper and lower bounds on the variables, consistently outperformed the other two direction distributions. Evidently $D3$ accelerates convergence of Hit-and-Run to its target distribution.

Cooling schedules $C1$ and $C5$ clearly performed better than the others. The good performance of $C1$ suggests that much can be gained from using an adaptive cooling schedule, however, since the optimum value would not be known in practice, a better estimator of the global maximum should be developed, since $C2$ did not perform as well as hoped. As Romeijn and Smith remarked, “The good performance of the simple cooling schedule $C5$ seems remarkable” ([140], page 118). As seen in Section 2, there is a theoretical reason for the success of IHR on positive definite quadratic programs, and although the test bed of problems included much broader global optimization test functions, perhaps the problems where $C5$ performed best shared similarities to this type of problem.

Another empirical test [140] investigated the effect of the number of dimensions on Hide-and-Seek with cooling schedules $C1$ and $C5$, where the problem dimension assumed values $n = 5, 10, 15, 20$, and 25. If Hide-and-Seek approximates annealing adaptive search and Improving Hit-and-Run approximates pure adaptive search successfully, then the number of observed record values (i.e., improving points) should grow linearly in dimension. The computational results strongly support this and linear regressions reported describe an excellent fit.

Computational experiments comparing Improving Hit-and-Run using the component-by-component reflection generator (see Section 4.1) with IHR using HD were reported in [141]. The sinusoidal function was used as a test problem for dimensions $n = 5, 10, 15$,and 20 and both centered and shifted ($C = 0°, 30°$) cases. In addition, the sinusoidal problem was discretized to test the step-function approach (discussed in Section 4.2). The discretized version of the problem kept the boundary constraints of $0° \leq x_i \leq 180°$ and additionally restricted the values to every 15°, $x_i \in \{0, 15, 30, 45, \ldots, 180\}$. Consequently four variations were run; IHR using HD and reflection on the continuous problem, and IHR using HD and reflection on the discrete problem with the step-function approach.

For the low dimensional problem ($n = 5$), the variations achieved nearly 100% success rate and there was little difference in the number of function evaluations. However for the higher dimensional problems ($n = 15, 20$), the generator with reflection clearly performed better than the generator using HD on both the continuous problem and using the step-function approach for the discretized problem.

The difference between performance on the continuous problem compared with performance on the discretized problem was overshadowed by the difference in generators. The trends were comparable between the continuous and discretized problems, indicating that the step-function approach is viable.

Neogi [116] conducted several numerical experiments using IHR and variations. He tested sixteen combinations of direction generator, cooling schedule, step-size and formulation. The two direction distributions were HD and cyclic CD. The two cooling schedules were $C5$ above (accept only improving points, $T_k = 0$), and $C3$ above (geometric, $T_k = 0.6^k$). Neogi also varied the step-size of the algorithm. In Hit-and-Run the step-size is determined by sampling uniformly along the line segment formed by intersecting the direction vector and the feasible region, which was called "full-range step-size." An alternative method is to adaptively restrict the feasible range of each variable to generate the step-size. This was called "adaptive step-size" and was motivated by [36] where the step-size could change to maintain a target acceptance ratio. See [116] for a detailed description of the adaptive step-size method. Neogi included in the numerical experiments two formulations with inequality constraints, a hierarchical formulation and a penalty function approach. These two formulations are described in Chapter 7.

Neogi used several test functions including the sinusoidal function, a fuel allocation problem, a 10-bar truss problem, and several composite design problems. The test functions other than the sinusoidal function are described in Chapter 7. On the sinusoidal test functions, the cyclic

CD direction generator performed better than HD, however the reverse was true for the fuel allocation problem. The penalty function approach generally performed better than the hierarchical approach. The differences between cooling schedules and step-size alternatives were small and seem problem dependent.

Computational studies are continuing to be conducted. Recently, Dereksdottir [42] compared discrete Hit-and-Run variations, and coupled IHR with multistart, which proved very effective. This is consistent with recent experiments [3] on continuous problems that indicate that the Hit-and-Run based algorithms make good improvement very quickly, but when they get trapped in a local optimum, a multi-start appears effective.

6. Summary

The use of Hit-and-Run as a generator for practical stochastic adaptive search algorithms is stimulated by the theoretical properties of pure adaptive search and adaptive search. One advantage to this approach is that fine-tuning of the generator is not required; it may be applied to any set. In addition, theoretical properties, including convergence to the global optimum with probability one, and some rate of convergence results, apply to a wide range of problems. The many variations of Hit-and-Run have been numerically tested on many test functions and successfully applied to real problems. Overall, the theoretical motivations and numerical experience lead us to believe that simulated annealing with Hit-and-Run is a promising approach to global optimization.

Chapter 7

ENGINEERING DESIGN APPLICATIONS

Applications of global optimization problems often arise when optimizing complex systems. Examples of such global optimization problems appear in engineering design where design problems are often highly nonlinear. The feasible region may be nonconvex and disconnected, and the functions may be nonconvex, nondifferentiable and even discontinuous. In addition, the decision variables may include both discrete and continuous variables. Many engineers are reporting success applying simulated annealing, genetic algorithms, evolutionary programming, hybrid methods, as well as variations of Improving Hit-and-Run and Hide-and-Seek to optimize complex engineering design problems.

Engineers always strive to optimize their designs — after all, who would not want the strongest, cheapest, and overall best design? Yet in practice, engineers tend to compare a very small number of designs due to the complex nature of the evaluation. Engineering design problems, structural design in particular, have used combined traditional gradient-based local optimization methods with multistart to tackle problems with multiple local optima [62, 171, 185]. While multistart has improved the effectiveness of local methods in practice [62], the application of stochastic adaptive search methods are also being used and providing practical solutions [115, 183, 184].

When optimizing complex systems, the functions used to evaluate the objective function and constraints are often termed "black-box" functions. This is because the functions are difficult to express in a closed form mathematical formula, but can be numerically evaluated through the use of computers. For instance, a function evaluation may involve a finite element analysis or a simulation. Hence the only information that can be derived from such a black-box function is a single function evaluation.

Using black-box functions allows engineers to easily change the analysis model depending on the application at hand. Many times the practicing engineer has tight deadlines and must respond quickly to changes in design. Consequently, they are primarily interested in any feasible designs, and secondarily interested in the optimal design. In addition, the problem definition changes frequently. For example, one day the optimization issue may be to minimize weight subject to buckling constraints and the next day may be to maximize stiffness under weight constraints. Similarly, an engineer designing a composite wing may want to use a different analysis to evaluate mechanical performance than to design the fuselage. Yet the engineer would prefer to use the same optimization algorithm, rather than to tailor the algorithm to the specific functions. Hence, one desires a versatile algorithm that can be applied to many kinds of functions.

Another use of an optimization algorithm during the design process is to explore the effect of discretizing a decision. For example, in a typical ten-bar truss optimization problem, the design variables are the diameters of the bars and the objective is to minimize the weight of the structure. The optimal solution for this problem depends on whether the diameters are assumed to be continuous variables, or are restricted to discrete values that are commonly available. This is an example of how manufacturing considerations can motivate discrete variables instead of continuous variables. It also illustrates the need for a robust optimization algorithm that can handle continuous and/or discrete variables.

The optimization process may be most effective early in the design process, when decisions are made that determine a large percentage of the final cost of the product. This adds to the need for solving many different problems and developing insight into the design space. Contrast this environment with a scheduling problem which may be solved repeatedly a hundred times a day but with slightly different data. This different environment calls for a different approach to optimization. Instead of fine tuning a large scale integer linear programming algorithm, one needs a flexible approach that is robust for a variety of functions.

This chapter focuses on formulating and solving global optimization problems that arise in engineering. However, the use of these algorithms is by no means limited to engineering design optimization problems. For example, Newton, et al. [120] have used the Improving Hit-and-Run algorithm to optimize a maximum likelihood function that was based on an application in hematology.

The first section discusses formulation issues including a penalty function approach for nonlinear nonconvex constraints. Section 2 presents a fuel allocation problem and Section 3 describes a three-bar truss problem with extensions to a ten-bar truss problem. Then Section 4 describes several optimization formulations for the design of polymer composite aircraft. The Hit-and-Run based optimization strategies presented in Chapter 6 have been successfully applied to optimize the design of wide body aircraft fuselage panels [58, 59, 182] with polymer composite materials. Through the collective efforts of Boeing, NASA, the University of Washington and others, a preliminary design software package called COSTADE (Cost/Composite Optimization Software for Transport Aircraft Design Evaluation) has been developed [105, 106] using Hit-and-Run based algorithms. The COSTADE design tool is aimed at designing laminated composite panels of various types subject to multiple loads and performance constraints.

1. Formulating Global Optimization Problems

In Chapter 1, optimization problems are classified according to the problem domain and whether the variables are continuous or discrete. Here, we are interested in global optimization problems with both continuous and discrete variables. The problems are typically constrained with many nonlinear inequalities and even finding a feasible solution is sometimes difficult.

The application of the algorithms in Chapter 6 is straightforward for optimization formulations with simple upper and lower bounds on the variables, however additional steps must be taken for the algorithms to handle the functional constraints. Global optimization problems arising in practice often have functional constraints imposed on their design variables and may take the form,

$$\begin{array}{lll} \text{minimize} & f(x) & \\ \text{subject to} & g_j(x) \geq 0 & j = 1, \ldots, m \\ & x_i^l \leq x_i \leq x_i^u & i = 1, \ldots, n \end{array}$$

where $g_j(x)$ are functional constraints, and x_i^l and x_i^u are lower and upper bounds on the design variables. Two approaches are presented to incorporate nonlinear nonconvex constraints in the optimization procedure: a hierarchical formulation and a penalty formulation.

1.1 Hierarchical Formulation

The hierarchical formulation is an approach that institutes a hierarchy of objective functions. One objective function may be to minimize infeasibilities and the other objective function may be the original $f(x)$.

This is similar to goal programming, or a Phase I/Phase II approach in linear programming [113], and can be extended to include a hierarchy of multiple goals and objectives.

A Phase I approach to find a feasible solution creates a modified optimization problem ($\mathcal{P}_1$) (Equation 7.1) that involves only the constraints of the original problem and not the objective function. The goal of this first phase of the optimization is to obtain a feasible solution. The following formulation maximizes the minimum of all constraint values,

$$(\mathcal{P}_1) \quad \begin{array}{lll} \text{maximize} & f_g(x) = \min_{j=1,\dots,n} \{g_j(x)\} & \\ \text{subject to} & x_i^l \leq x_i \leq x_i^u & i = 1, \dots, n. \end{array} \tag{7.1}$$

If x is infeasible, then $f_g(x) < 0$ and the value of $f_g(x)$ reflects the error in constraint satisfaction by measuring the worst constraint value.

The hierarchical formulation was originally developed in the context of structural optimization [116]. Here, the primary goal is to minimize the weight of the structure, but performance measures based on mechanics, such as strength, must exceed a certain minimum threshold. Once the threshold is satisfied it may still be desirable to increase that performance measure. For example, the strength of the structure must exceed a certain level to be considered feasible, but a stronger structure is even better.

The optimization problem ($\mathcal{P}_2$) is defined by a hierarchy of objective functions $f(x)$ and $f_g(x)$ as follows,

$$(\mathcal{P}_2) \quad \begin{array}{llll} \text{minimize} & f(x) & (\textit{primary}) \\ \text{maximize} & f_g(x) & (\textit{secondary}) \\ \text{subject to} & g_j(x) \geq 0 & j = 1, \dots, m \\ & x_i^l \leq x_i \leq x_i^u & i = 1, \dots, n \end{array} \tag{7.2}$$

where the primary objective function $f(x)$ is considered for minimization first. However, if $f(x)$ remains at the same value for any two candidate solutions during the process, then the secondary objective of maximizing $f_g(x)$ is considered, while maintaining feasibility. In the case of structural design of composite laminates, this hierarchical formulation is used to minimize the overall weight of the structure while secondarily maximizing the performance of the structure. Some of the design variables in this problem represent fiber angles of composite plies, and other design variables represent the number of plies (see detailed description in Section 4). While the number of plies affects weight, the fiber angles affect only performance. Hence this hierarchical formulation allows an engineer to simultaneously minimize weight and maximize performance while ensuring a feasible level of performance. The optimization algorithms presented in Chapter 6 are well-suited for this type of formulation because they need only function evaluations, and can easily incorporate a hierarchical comparison of solutions.

1.2 Penalty Formulation

The penalty function method transforms a constrained optimization problem to an unconstrained problem by imposing a penalty for any constraint violation. There are several traditional approaches such as the exterior penalty function method, the interior penalty function method and the extended penalty function method [13]. In all cases, the penalty value is sequentially updated in a series of optimization runs which lead to the optimal constrained design. This approach belongs to the classical *sequential unconstrained minimization technique* or SUMT [170]. Of the several penalty function approaches, the exterior penalty function method is straightforward to implement in the context of random search methods. Typically a large fixed penalty is imposed for such implementations.

The application of the exterior penalty on the violated constraints results in the following problem formulation ($\mathcal{P}_3$),

$$(\mathcal{P}_3) \quad \begin{array}{ll} \text{minimize} & f(x) \; + \; r_p \, \varepsilon(x) \\ \text{subject to} & x_i^l \leq x_i \leq x_i^u \qquad i = 1, \ldots, n \end{array}$$

where $r_p \geq 0$ is a penalty factor, and $\varepsilon(x)$ is defined as follows,

$$\varepsilon(x) \;=\; \sum_{j=1}^{m} \min\{0, g_j(x)\}^2 \tag{7.3}$$

for m constraints $g_j(x)$, $j = 1, \ldots, m$.

The $\varepsilon(x)$ is a measure of the violation of the constraints. If the constraints are all satisfied, then $\varepsilon(x) = 0$ and hence the minimum of ($\mathcal{P}_3$) is also the minimum of the original constrained problem. In the penalty formulation, a squared sum of all the violated constraint values is chosen to be the error measure for $\varepsilon(x)$. This error measure captures the violations of all the constraints rather than just the worst case as in the maxi-min expression in Equation 7.1.

The factor r_p is a variable penalty factor. The effect of r_p on the optimization process is as follows. For small values of r_p, the minimum of $f(x) \, + \, r_p \, \varepsilon(x)$ results in major constraint violations for the original minimization problem because the value of the objective function may dominate the penalty term. On the other hand, an increasing value of r_p causes greater penalization of infeasibility and forces the constraints to be satisfied. In the limit as $r_p \to \infty$, a feasible design is ensured. However, previous applications of random search methods have found that a large value of r_p was sufficient to generate feasible designs. Studies in [116, 162] demonstrate the benefit of introducing a penalty schedule analogous to a cooling schedule in simulated annealing.

Penalty Schedule: As described here, the penalty function approach uses a penalty schedule to determine r_p, similar to the temperature cooling schedule in simulated annealing. The penalty factor is gradually increased as follows,

$$r_{p+1} = Cr_p \tag{7.4}$$

where C is a constant greater than 1. The study in [116] reports that the golden section ratio of 1.618033 worked well. The initial value of r_p is chosen to be a small value, empirically set to 0.1% of the ratio between $f(x)$ and $g(x)$ values. The penalty factor is increased at the end of each temperature step until it reaches a large number ($\sim 10^8$), at which point it is held constant. Neogi [116] describes a comparative study of fixed versus variable penalty factors and the results demonstrated the effectiveness of a variable penalty factor using the penalty schedule.

A temperature dependent penalty factor is used by Stern in [162],who solves a block angular form reduction problem of a matrix by combinatorial simulated annealing. It is shown that a temperature dependent penalty function approach outperforms the standard simulated annealing approach.

Consider further the analogy between temperature cooling schedule and penalty factor schedule. When temperature is high in simulated annealing, the algorithm has a high probability of accepting a non-improving point. Consequently, it might escape a local optimum in favor of finding a better one. By the same token, when the penalty factor is low, the algorithm can accept an infeasible point and travel through an infeasible space to find a better portion of the feasible region. Because the feasible region may be nonconvex, it can be advantageous to accept an infeasible point in the beginning of the run. In addition, cooling the temperature to zero is analogous to increasing the penalty factor to infinity.

Neogi [116] also performs a numerical comparison of the hierarchical approach versus the penalty approach with a penalty schedule. He concludes that the penalty approach worked better than the hierarchical approach in most situations.

2. Fuel Allocation Problem

Global optimization has many relevant applications in chemical engineering practice where problems are often nonlinear and constrained

[51, 175]. Hovanessian and Stout [75], in 1963, describe a fuel allocation problem as an economic optimization problem for a power plant consisting of a number of boiler-turbine-generator combinations each with specified efficiency-load-fuel characteristics. The objective of the problem is to minimize the fuel cost and find the operating level as well as the fuel mixture ratio of each boiler-turbine-generator combination. The problem reappears in [74], and is used as an example by Luus and Jaakola [104] and Cardosa, et al. [31] to test their optimization algorithms.

The problem, summarized by Luus and Jaakola [104], is to minimize the fuel oil purchased to produce 50 mega-watts (MW) from a two-boiler-turbine-generator combination that can use fuel oil, blast furnace gas (BFG), or any combination of these two sources of fuel. The fuel allocation problem is

$$
\begin{aligned}
\text{minimize} \quad & F = f_1 x_3 + g_1 x_4 \\
& \text{where } f_1 = 1.4609 + 0.15186 x_1 + 0.00145 x_1^2 \\
& \text{and} \quad g_1 = 0.8008 + 0.2031 x_2 + 0.000916 x_2^2 \\
\text{subject to} \quad & (1 - x_3) f_2 + (1 - x_4) g_2 \leq 10 \\
& \text{where } f_2 = 1.5742 + 0.1631 x_1 + 0.001358 x_1^2 \\
& \text{and} \quad g_2 = 0.7266 + 0.2256 x_2 + 0.000778 x_2^2 \\
& x_2 = 50 - x_3 \\
& 18 \leq x_1 \leq 30 \\
& 14 \leq x_2 \leq 25 \\
& 0 \leq x_3 \leq 1 \\
& 0 \leq x_4 \leq 1.
\end{aligned}
$$

The variables x_1 and x_2 are the mega-watts produced by generators 1 and 2, respectively. The fraction of fuel oil used in generators 1 and 2 are denoted by x_3 and x_4, respectively. The fraction of blast furnace gas used in each generator is determined by $(1 - x_1)$ and $(1 - x_2)$. The equality constraint, $x_2 = 50 - x_3$, ensures that 50 MW of power is produced.

The fuel oil requirement in tons per hour at generators 1 and 2 is given by the functions f_1 and g_1. The blast furnace gas in fuel units per hour at generators 1 and 2 are represented by f_2 and g_2. The first functional constraint, involving f_2 and g_2, indicates the limited availability of blast furnace gas. Operating ranges are given as lower and upper bounds on the variables. The global optimum to the fuel allocation problem is: $x_1 = 30$, $x_2 = 20$, $x_3 = 0$, $x_4 = 0.58366$, with $F = 3.0521$.

This fuel allocation problem has been used to test several different algorithms. Luus and Jaakola [104] apply a method that systematically reduces the size of the search region using direct search to many problems including this fuel allocation problem. Cardosa, et al. [31] compare a simulated annealing variant called non-equilibrium simplex-simulated annealing algorithm with a few other algorithms on the fuel allocation problem. Neogi [116] reports results of different versions of Hit-and-Run on the fuel allocation problem, and shows that the variations were robust efficient methods. Neogi [116] observes that the IHR variations with a penalty formulation compared favorably to the non-equilibrium simplex-simulated annealing algorithm on the tests performed. This small fuel allocation problem is just a sample of simple practical problems.

3. Truss Design Problem

Truss design is a common engineering problem for both civil engineering and mechanical engineering. A truss is composed of two or more truss members, which are joined at nodes. Problems of this type are often *statically indeterminate*, that is, indeterminate within the framework of statics. For example, the internal and reaction forces among the members of the truss are not determined by simply drawing a free-body diagram of the members and writing the corresponding equilibrium equations. To solve the problem, deformations obtained when the geometry of the problem is considered must complement the equilibrium equations. This makes the truss design problem a black-box optimization problem.

3.1 Three-Bar Truss Design

To demonstrate how a truss design problem is formulated, consider a simple three-bar truss example. This problem has closed form load expressions and can lead to an understanding of more complicated truss structure problems. The objective when optimizing the design of a simple three-bar truss is to minimize the weight of the structure while supporting a total load P acting vertically downward. The geometry of the problem is shown in Figure 7.1.

The internal loads P_1, P_2, and P_3 in the corresponding members in the figure are,

$$P_1 = \frac{b_1 b_3}{\sqrt{2} b_1 b_3 + b_1 b_2 + b_2 b_3} P$$

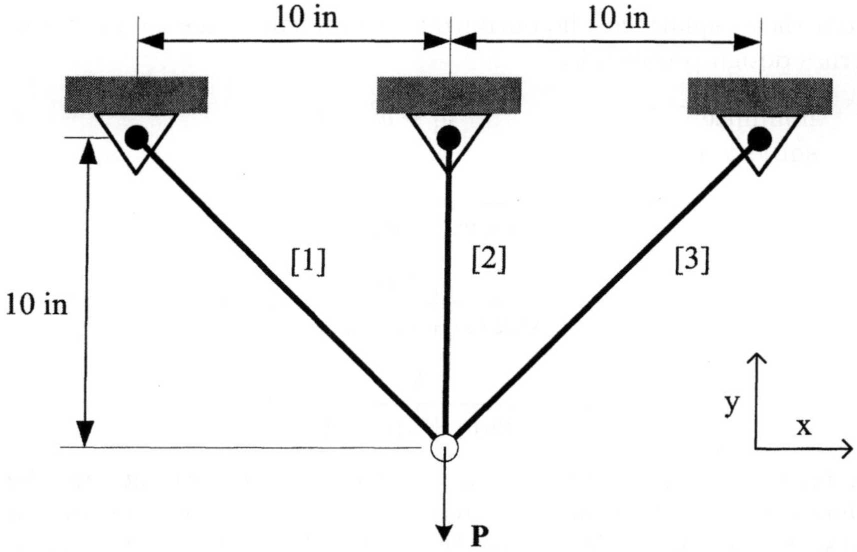

Figure 7.1. Three-bar truss diagram.

$$P_2 = \frac{(b_1 + b_3)b_2}{\sqrt{2}b_1b_3 + b_1b_2 + b_2b_3}P \tag{7.5}$$
$$P_3 = \frac{b_1b_3}{\sqrt{2}b_1b_3 + b_1b_2 + b_2b_3}P$$

where b_1, b_2, and b_3 are the cross-sectional areas of the truss members 1, 2, and 3, respectively. These loads are derived from the stress-displacement equations detailed in Haug and Arora [66, Example 1.2]. The expressions for the internal loads of members 1 and 3 are the same due to the symmetry of the problem. It is also important to realize that the external load P is supported by all the three truss members and changing a cross-sectional area of one member affects the load distribution to all the members.

The axial stress σ in a truss member is calculated by dividing the internal load of the member by its cross-sectional area. Hence, the axial stresses of members 1, 2, and 3, are,

$$\sigma_1 = \frac{P_1}{b_1}, \quad \sigma_2 = \frac{P_2}{b_2}, \quad \sigma_3 = \frac{P_3}{b_3}, \tag{7.6}$$

respectively.

From these equations, the optimization problem of the simple three-bar truss design problem is,

$$
\begin{array}{lrcl}
\text{minimize} & & b_1^2 + b_2^2 + b_3^2 & \\
\text{subject to} & & & \\
& \sigma_1 & = \dfrac{b_3}{\sqrt{2}b_1b_3 + b_1b_2 + b_2b_3}P & \leq \sigma_{max} \\
& \sigma_2 & = \dfrac{b_1 + b_3}{\sqrt{2}b_1b_3 + b_1b_2 + b_2b_3}P & \leq \sigma_{max} \\
& \sigma_3 & = \dfrac{b_1}{\sqrt{2}b_1b_3 + b_1b_2 + b_2b_3}P & \leq \sigma_{max}
\end{array}
$$

where b_1, b_2, and b_3 are the design variables. The constraints are the maximum axial stresses in the three truss members. The maximum stresses, which usually depend on the material property of the members, ensure that the truss does not fail when the load P is imposed. The quadratic objective function is chosen here because the linear version produces zero cross-sectional area for members 1 and 3. When the load $P = 5000$ lbs and $\sigma_{max} = 340$ psi, the optimal solution found is $(b_1, b_2, b_3) = (4.159, 11.766, 4.159)$ in^2. Neogi [116] has used this simple three-bar truss problem to illustrate the dependence between the design variables (cross-sectional areas b_1, b_2, and b_3) and the stresses.

A more general case of this three-bar truss design problem is when the load P has positive components in both vertical and horizontal directions. The problem becomes more complicated and requires computation of vertical and horizontal displacements as detailed in [66, Example 1.2]. Displacements must be considered in a realistic design problem. To compute the displacements, a *stiffness method* is employed that calculates the node-wise stiffness of the structure. More details of the stiffness method can be found in structures textbooks such as [70] and [76].

3.2 Ten-Bar Truss Design

The ten-bar truss design example is another example of a truss design problem. This problem is considered a medium size problem with six nodes, ten members, and vertical and horizontal displacements. This particular truss problem has been used extensively in the literature to evaluate optimization algorithms [18, 66, 116].

The objective of the ten-bar truss problem is to minimize the overall weight of the structure. The geometry for a specific ten-bar truss is shown in Figure 7.2. There are two sets of constraints. The first set involves the axial stresses σ in the members similar to those in the three-bar truss problem. The second set limits the maximum displacements δ of any node of the truss. The optimization problem is formally stated as follows,

$$\begin{array}{llll} \text{minimize} & f_1 & = & \sum_{i=1}^{10} \rho A_i l_i \\ \text{subject to} & |\sigma_i| \leq \sigma_{all} & i = 1, \ldots, 10 & \\ & \delta_{jx}, \delta_{jy} \leq \pm 2 & j = 1, \ldots, 4 & \end{array}$$

where A_i and l_i are the cross-sectional areas and the lengths of the ten members, f_1 is the weight of the truss, ρ is a constant representing the density of the material, σ_i is the stress level in each member i, and δ_{jx}, δ_{jy} are the displacements in the x and y directions for each free node j. In the ten-bar truss diagram in Figure 7.2, there are ten members and four free nodes, nodes 1 to 4. Nodes 5 and 6 are not free because they are hinged to the wall and hence have zero displacement. The allowable stress for all members is specified by σ_{all} which is 25,000 psi in both tension and compression. The allowable displacement of all free nodes in x and y directions is limited to ± 2.0 in. The loading condition is shown in the figure. The Young's modulus of elasticity is $E = 10^7$ psi and the material density is $\rho = 0.1$ lb/in^3. Range limits on the areas are $A_i^{min} = 0.1$ in^2 and $A_i^{max} = 33.5$ in^2.

The design variables in the ten-bar truss problem are the cross-sectional areas A_i, and the lengths l_i of the ten members. The lengths are considered to be continuous variables, but cross-sectional areas are first considered to be continuous variables, and later to be discrete variables to reflect standard diameters of members. The design problem is more realistic when restricting the diameters to a discrete set of standard values. For the mixed continuous/discrete version of the ten-bar truss problem, the cross sectional areas A_i are restricted to a discrete set of 42 values given as {1.62, 1.8, 1.99, 2.13, 2.38, 2.62, 2.63, 2.88, 2.93, 3.09, 3.13, 3.38, 3.47, 3.55, 3.63, 3.84, 3.87, 3.88, 4.18, 4.22, 4.49, 4.59, 4.8, 4.97, 5.12, 5.74, 7.22, 7.97, 11.5, 13.5, 13.9, 14.2, 15.5, 16.0, 16.9, 18.8, 19.9, 22.0, 22.9, 26.5, 30.0, 33.5} in^2.

Fourteen variations of Hit-and-Run based algorithms have been applied to the ten-bar truss problem, and complete numerical results are presented in [116]. To summarize briefly, the results on the ten-bar truss problem show that the penalty function approach (described in Chapter 6 Section 1.2) performs better than a formulation with strict constraints. For the mixed continuous/discrete version of the ten-bar truss problem, the version of the algorithm using HD is more effective, in terms of success percentage, than the version using CD. When the

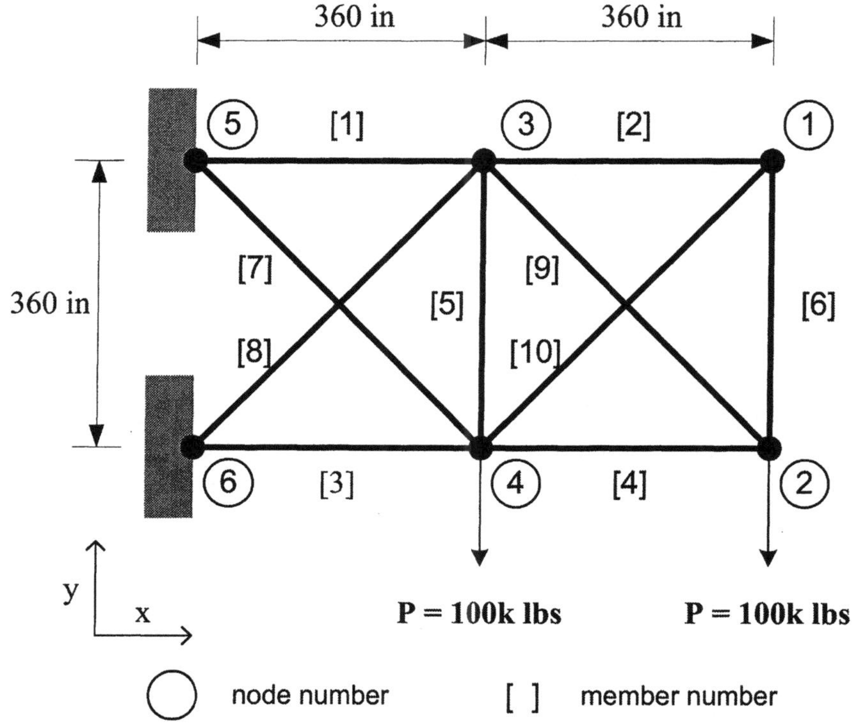

Figure 7.2. Ten-bar truss diagram.

cross-sectional areas are treated as continuous variables, HD and CD perform similarly. The minimum weight design found in the continuous version is 5037.3 lbs, and in the mixed continuous/discrete version is 5466.9 lbs, which improves upon the minimum weight design that is reported in [18] using a simulated annealing implementation. This example also illustrates the type of study a design engineer may wish to perform, comparing weight savings if using non-standard cross-sectional areas.

4. Optimal Design of Composite Structures

Composite structures combine multiple layers of new fiber-reinforced composite materials. Such structures can save weight relative to the same structure made of conventional materials such as aluminum. This weight saving advantage becomes particularly beneficial in weight-sensitive applications such as in aerospace structures.

In addition to the high strength-to-weight and stiffness-to-weight ratios, composite materials also allow engineers to *design the material* as well as the structure by *tailoring* the directional stiffness and strength as required for the structure [58, 59, 174]. On one hand, this can be viewed as an advantage because it offers added flexibility in the design and the ability to be more precise in requirements. On the other hand, designing the material and the structure simultaneously increases the burden to engineers due to the fact that the number of design variables increases dramatically. The interactions between the design variables also amplify the complexity of the design process and result in requiring more computational effort. The optimal design of composite structures is an active research area [9, 10, 62, 184] and the capability of today's computer technology and global optimization algorithms enable design engineers to explore the interaction amongst the design variables with respect to overall system performance and objectives. It also allows the engineer to evaluate tradeoffs between various objectives, including weight and performance.

Large composite structures such as aircraft are typically composed of a collection of composite panels. These panels are commonly designed based on lamination theory [78]. Unlike conventional engineering materials which are often assumed homogeneous, each of these panels are composed of individual *laminae* or *plies*, stacked on top of each other and bonded together. The bonded assembly is called *laminate.* For example, a graphite epoxy ply consists of graphite fibers embedded in an epoxy resin matrix. Each ply contains either unidirectional or woven fibers. To simplify our discussion, we concentrate on unidirectional fiber plies throughout this section. The specific order and fiber orientations of the plies in a laminate is referred to as the *stacking sequence.* Lamination theory relates the complex interactions between these plies to predict mechanical properties such as stiffness and strength of the overall laminate.

Figure 7.3 shows a simple composite laminate consisting of three unidirectional fiber plies. The laminate in Figure 7.3 also illustrates that the orientation of the fibers in one ply may differ from the orientation of another ply. By convention, the fiber orientation θ associated with a given ply i is identified using a subscript. Thus, the fiber angle in ply 1 is denoted θ_1, the fiber angle in ply 2 is denoted θ_2, etc. In a composite panel, the orientation of the fiber angles together with the number of plies determine the performance of the structure.

For conventional engineering materials, such as steel, a thicker structure naturally implies a stronger structure. This implication is not necessarily true for a composite structure because of the strong influence of

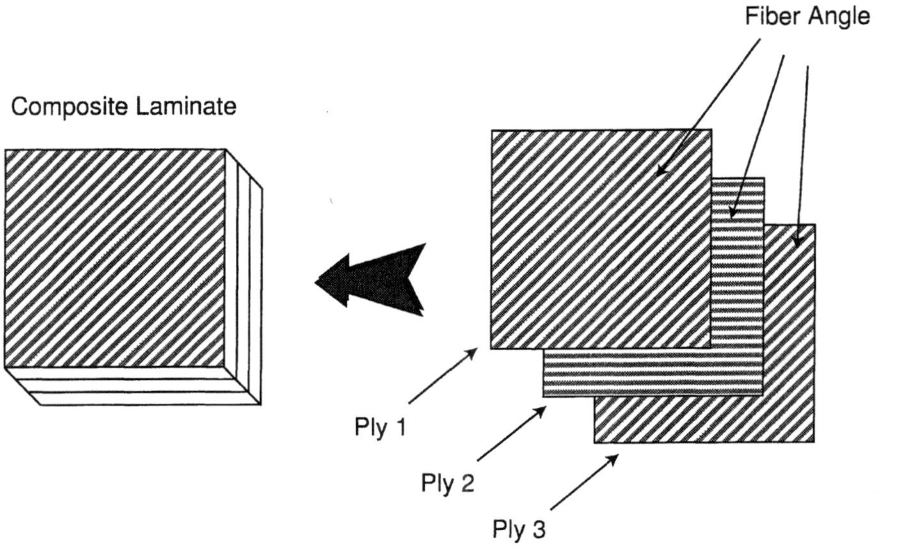

Figure 7.3. A 3-ply composite laminate with variation in fiber angle from one ply to the next.

fiber directions. In fact, the mechanical performance of any composite structure depends so heavily on the fiber directions of the laminate that a thinner laminate with appropriate fiber angles is often superior to a thicker one with fiber angles in sub-optimal directions.

In the following section, we present several examples of how the algorithms presented in Chapter 6 can be used to design composite structures. The application described here has been implemented in an optimization software package for the design of large laminated composite panels to be used in aircraft fuselage [58, 59, 105]. Optimization has been shown to be a useful tool in the design process, allowing engineers to quickly find a usable solution, explore the design space, and perform tradeoff studies.

4.1 Optimal Design of a Composite Stiffened Panel

Consider the design of a stiffened panel made from advanced composite materials, such as graphite epoxy. An example of a stiffened panel is illustrated in Figure 7.4. The design variables for this particular type of structure include the fiber angles associated with each ply in the skin $\theta_{i,\text{skin}}$ and stiffener $\theta_{i,\text{stiffener}}$, the number of plies in the skin n_{skin} and in the

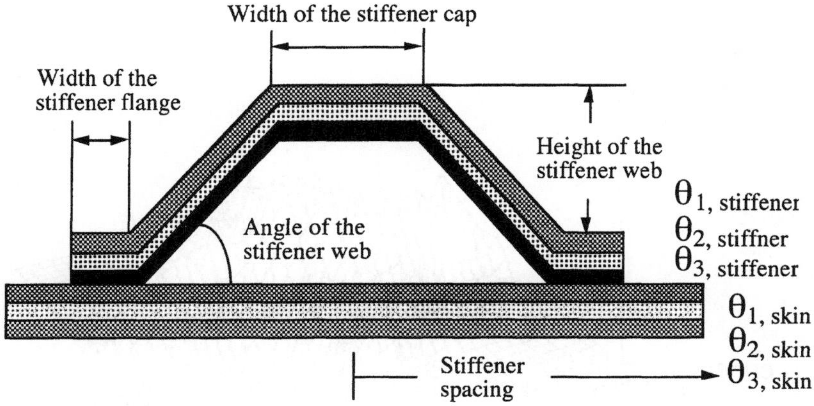

Figure 7.4. Design variables in a composite stiffened panel.

stiffener $n_{\text{stiffener}}$, and the geometry variables for the stiffener (width of cap, width of flange, height of web, angle of web, and stiffener spacing). Some of these variables are continuous and others are discrete. For instance the fiber angles may take on continuous values between +90 and −90 degrees, or may be restricted to discrete values, such as 0, ±45, or ±90 degrees. In the formulation for this structure, binary variables t_i, for $i = 1, \ldots, n$ are used to indicate whether ply i exists in the laminate, and n is viewed as an upper bound on the number of plies.

The constraints and/or the objective function can be described as black-box functions where the functions often are available only in the form of a computer subroutine. For example, stiffness of a composite laminate can be calculated using classical lamination theory [78], or a finite element analysis can be used to evaluate stiffness.

To give an example of the global nature of the equations used for mechanics analysis, a plot of the in-plane stiffness of a four ply, symmetric laminate, $[\theta_1, \theta_2, \theta_2, \theta_1]$, using classical lamination theory, is shown in Figure 7.5. The in-plane stiffness is one of many relevant performance measures that engineers might use. The greatest in-plane stiffness in the laminate $[\theta_1, \theta_2, \theta_2, \theta_1]$, for the particular load condition applied, occurs when the fiber angles are at 0 degrees, as illustrated in Figure 7.5. The plateau in the graph represents infeasible designs, with a stiffness that is less than a prescribed critical value. If stiffness is used as an objective function, it can be seen to be nonlinear and nonconvex. If stiffness is used in the constraints to allow only those designs above the threshold of critical stiffness, the feasible region itself is nonconvex and even has holes in it. This illustrates that even attaining feasibility can be a difficult problem.

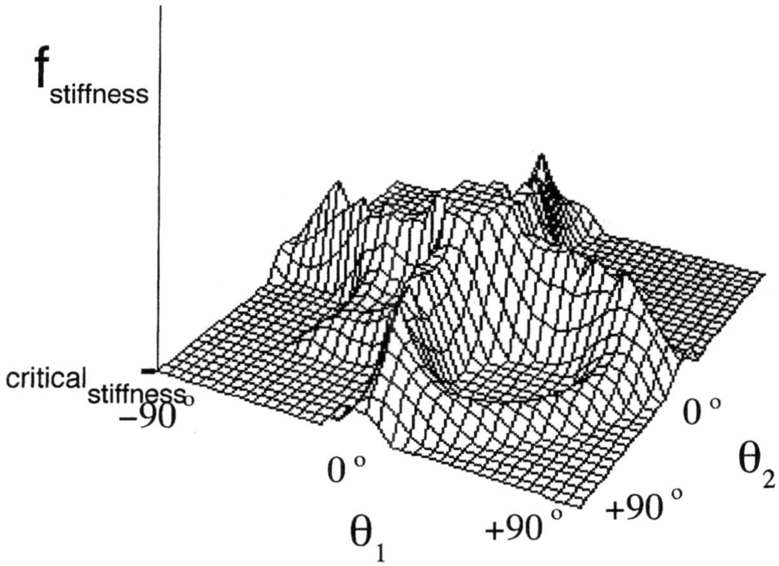

Figure 7.5. Graph of in-plane stiffness for a four ply, symmetric laminate, $[\theta_1, \theta_2, \theta_2, \theta_1]$.

Basic Formulation: The basic formulation for the composites design problem can be written as

$$\begin{array}{lll} \text{minimize} & f(x) & \\ \text{subject to} & g_j(x) \geq 0 & \text{for } j = 1, \ldots, m \\ & L_i \leq x_i \leq U_i & \text{for } i = 1, \ldots, n \\ & x \in X & \end{array} \tag{7.7}$$

where the objective function $f(x)$ is a measure of the performance and/or efficiency of the structure, often weight or cost of the structure. The m constraints ensure that margins of safety $g_j(x)$ based on mechanics calculations are positive for given loading conditions. Upper and lower bounds, $L_i \leq x_i \leq U_i$, provide box constraints on the design variables. The design variables can be continuous or discrete. A set of discrete values can be specified in the set X.

As with any structure, composite structures are designed to perform under different criteria depending upon the functional tasks. The most common requirements are load requirements. In order to meet these load requirements, the overall composite structure relies upon the properties of the individual plies. The inequality constraints $g_j(x) \geq 0$ typically address the mechanics issues of the overall structure which require non-negative margins of safety [105].

The margin of safety (MS) for a load dependent analysis is defined as,

$$g_j(x) = \mathrm{MS}_j(x) \;=\; \frac{F_j^{allow}}{F_j^{max}(x)} - 1 \tag{7.8}$$

where j indicates an analysis criterion, $F_j^{max}(x)$ is the maximum value attained by the candidate solution for a given load (computed using the appropriate analysis model as a black-box function). The allowable value F_j^{allow} is supplied by the engineer and depends on the material being used. If the maximum value is not greater than the allowed value for analysis criterion j, $F_j^{max}(x) \leq F_j^{allow}$, then the proposed design x is feasible with $\mathrm{MS}_j(x) \geq 0$. If this is not satisfied, then the margin of safety is negative which means that the proposed structure does not withstand the given loads. Possible load dependent analyses such as *strain* and *minimum buckling load* can be converted to margins of safety in this manner.

For load independent analysis criteria such as *strength* or *stiffness*, the margin of safety is defined as,

$$g_j(x) = \mathrm{MS}_j(x) \;=\; \frac{F_j^{actual}(x)}{F_j^{req}} - 1 \tag{7.9}$$

where $F_j^{actual}(x)$ is obtained from the structural analysis j and F_j^{req} is the requirement prescribed by the engineer. The design x is considered feasible with respect to analysis j if $F_j^{actual}(x) \geq F^{req}$. For example, in Equation (7.9), if the required stiffness is fixed at 100,000 lbs/in and the current design stiffness is 120,000 lbs/in, then the margin of safety is 0.2. For any analysis criterion, a negative margin of safety indicates an unacceptable design.

There are many possible analysis modules engineers may consider when designing a composite structure. In COSTADE [105, 107], the following analysis criteria are available - *strength, stiffness, buckling load, strain, Poisson ratio mismatch, damage tolerance, facesheet wrinkling and dimpling, flexural core crush, crimping* and *warpage.* Converting these criteria to margins of safety provides a standardized way to describe feasibility of a design.

When multiple analysis criteria are chosen simultaneously, there are many constraints that make it difficult to find a feasible design. As discussed in Sections 1.1 and 1.2, a hierarchical or penalty function approach can be used. A hierarchical formulation can maximize the worst margin of safety so that a very large margin of safety for one criterion does not overshadow a small negative value of a different criterion. This ensures $g_j(X) \geq 0$ for each criterion. For the penalty formulation, a

E_1	132.5 GPa	ε_1^{allow}	0.0143
E_2	9.4 GPa	ε_2^{allow}	0.0060
G_{12}	4.9 GPa	γ_{12}^{allow}	0.0100
Density	1.02 g/mm^3	Y^T	56 MPa
Ply thickness	0.1880 mm	S	49 MPa

Table 7.1. Assumed ply material properties for AS4/3501-6 graphite epoxy.

Load condition	N_x (lbs/in)	N_y (lbs/in)	N_{xy} (lbs/in)	M_x (lbs/in)
1	2000	1000	1000	2000
2	2000	1000	-1000	2000

Table 7.2. Loading conditions.

squared sum of margins of safety is used to minimize the total of the violated constraints. Numerical experiments performed on test problems in [116] indicate that the penalty function approach finds a feasible design more often than the hierarchical approach.

Example design: Graesser, et al. [59] presents a design methodology for a laminated composite stiffened panel, subjected to multiple in-plane loads and bending moments. The panel optimized contains a hat stiffener as shown in Figure 7.4.

For this example, the optimization formulation follows Equation 7.7 where the objective function $f(x)$ is the weight of the panel. The design variables are the fiber angles in the skin and stiffener and the geometry variables that describe the stiffener. The constraints represent the maximum strain criteria. The ply material is assumed to be AS4/3501-6 graphite epoxy, with material properties listed in Table 7.1.

We seek the lightest possible panel design that can support the two loading conditions listed in Table 7.2 with constraints that no ply strain can exceed the specified maximum strain under either loading condition. The analysis model is based on classical lamination theory and is described in detail in [59]. For the discussion here, we assume that strain is calculated with a subroutine as a black-box function.

The optimization process in [59] uses the hierarchical formulation described earlier for a symmetric laminate with a hat-type stiffener. They optimize hat-stiffened panels consisting of an eight-ply symmetric skin laminate reinforced with symmetric 8-, 10-, 12- and 14-ply stiffeners. A

Skin laminate	Stiffener laminate	Stiffener spacing (in)	Weight (lbs/in^2)
$[37/-49/-1/81]_s$ (E_x=8.1 msi)	$[0]_{4s}$ (E_x=19.2 msi)	64	0.00379
$[39/85/-44/-1]_s$ (E_x=7.9 msi)	$[0]_{5s}$ (E_x=19.2 msi)	80	0.00379
$[87/40/-1/-43]_s$ (E_x=7.9 msi)	$[0]_{6s}$ (E_x=19.2 msi)	97	0.00379
$[-36/3/46/-80]_s$ (E_x=8.0 msi)	$[0]_{7s}$ (E_x=19.2 msi)	112	0.00379

In all of the above designs, the width of the stiffener flange was 1 in, the height of the stiffener web was 2 in, the width of the stiffener cap was 2 in, and the angle of the stiffener web was 90°.

Table 7.3. Minimum weight designs for the maximum strain design constraint.

symmetric laminate assumes the stacking sequence is symmetric about the mid-plane, thus reducing the number of fiber angle variables by half. Therefore, four distinct fiber angles are selected to define the skin laminate and four, five, six and seven fiber angles are selected to define the stiffener laminate. The upper and lower bounds on the geometry variables are: width of the stiffener flange between $1-2$ in, height of the stiffener cap between $1-2$ in, width of the stiffener cap between $1-2$ in, angle of the stiffener web between $45° - 90°$, and stiffener spacing between $10-200$ in.

The designs summarized in Table 7.3 have been found using Improving Hit-and-Run and its variations [59]. The subscript s to the solutions in the table indicates that the laminate is symmetric. The four designs have equal weight but different numbers of plies in the stiffener. The stiffener spacing increases as the number of plies in the stiffener increases, providing alternate designs of equal weight. All the fiber angles in the stiffener are oriented at 0° which is a direct result of the mechanics model. Because the stiffener is assumed to carry loads in the axial direction only, the optimizer selects a $[0]_n$ laminate in all cases, as is expected. All of the stiffener geometry variables are at their upper or lower bounds, as noted in the table.

Several other cases are considered in [59]. One analysis explores the effect of discretizing the fiber angles. A hat-stiffener panel where fiber angles are limited to multiples of 15° was 8% heavier in weight than when the fiber angles are allowed to take on continuous values.

The design methodology of this example can be used to design various types of composite structures. Other structures include a composite flat plate discussed in [58, 185], I-beams [145], sandwich panel [95], and more complex fuselage barrels [106, 116]. As the structure gets larger, the design problem becomes more complex. Such extensions to larger structures are discussed next.

4.2 Extensions to Larger Structures

Extending the optimization formulation for a specific cross-section to cover an entire panel introduces many complications. In actual practice, a large composite panel is rarely subjected to a uniform distribution of load. Instead, loading varies over the length and width of the panel, with some areas experiencing relatively high loads and others experiencing relatively low loads. For example, consider the keel panel in a commercial aircraft (i.e., the panel in the underbelly of aircraft fuselage). Typically, there is an opening in the keel panel to accommodate the landing gear. That portion of the keel panel just aft of the landing gear usually experiences relatively high loads, and the magnitude of loading gradually decreases at regions away from the gear. The engineers are typically provided with a "map" of predicted load contours over the length and width of the keel panel, as shown schematically in Figure 7.6. The smooth lines shown at internal regions of the panel represent lines of constant load intensity. These load contours are predicted during preliminary analysis and design of the overall aircraft, and the job of the engineers is to design a panel that can safely sustain these loads. In the case of a composite sandwich keel panel, this means that the engineers must specify the stacking sequence of both facesheets as well as the core material and thickness.

Of course, one solution is to identify a stacking sequence that can support the highest concentration of load, and then to use this stacking sequence throughout the entire length and width of the panel. While such a design is mechanically feasible, it is certain to be heavier than required, because those regions of the panel that experience relatively low loads are overdesigned. Fortunately, it is possible to change the stacking sequence used in a composite panel by "dropping" plies (i.e., discontinuing a ply) in regions of low loads. In the optimization model, the panel is divided into regions, as shown in Figure 7.7. The load contours shown in Figure 7.6 are used to infer the magnitudes of the loads applied to each region in Figure 7.7. The star symbol ★ in Figure 7.7 denotes the region that experiences the highest loading, and we call this region the *key* region.

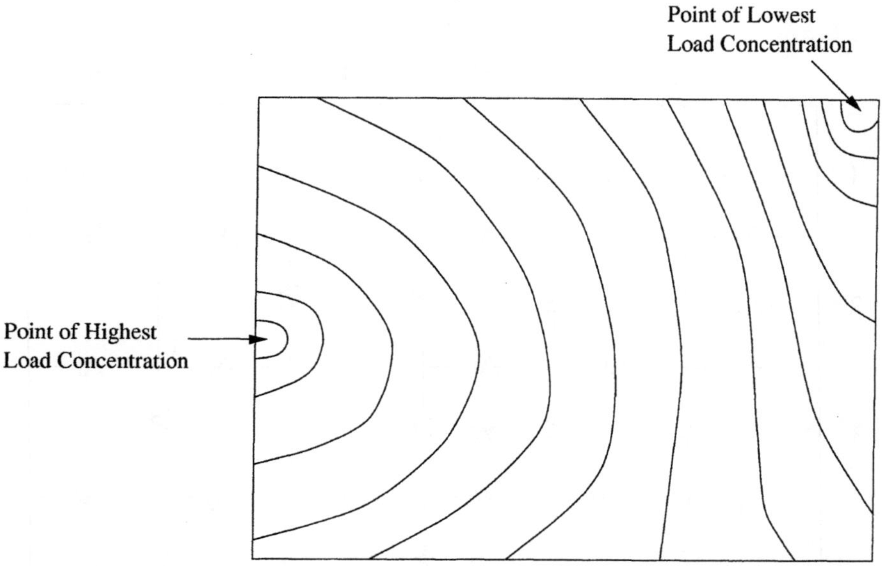

Figure 7.6. Non-uniform loading of a large composite panel.

In order to avoid overdesigning the panel, an optimization approach must tailor the design of specific regions to the anticipated loads. However, the number of plies and fiber orientations cannot vary too dramatically or else the panel becomes impractical to manufacture. From the manufacturing view point, it is preferred to have a common stacking sequence over as many regions as possible, and then reduce the number of plies gradually between adjacent regions. The engineers must avoid abrupt changes in fiber orientation within a ply to eliminate seams that are mechanically weak. A mechanism of allowing a ply to continue through adjacent regions until it is no longer required is called *blending*. The optimization approach in [95] uses blending rules to achieve a blended panel that is not only manufacturable, but also tailored to support the load concentrations.

Given the load conditions, the key region of the panel can be identified. Because the key region undergoes the largest load concentration, it is normally the region that requires the most number of plies. Starting from the key region, the number of plies in adjacent regions can be dropped as long as the required stiffness and strengths are maintained. Once a ply is dropped, it cannot be added back to the stacking sequence

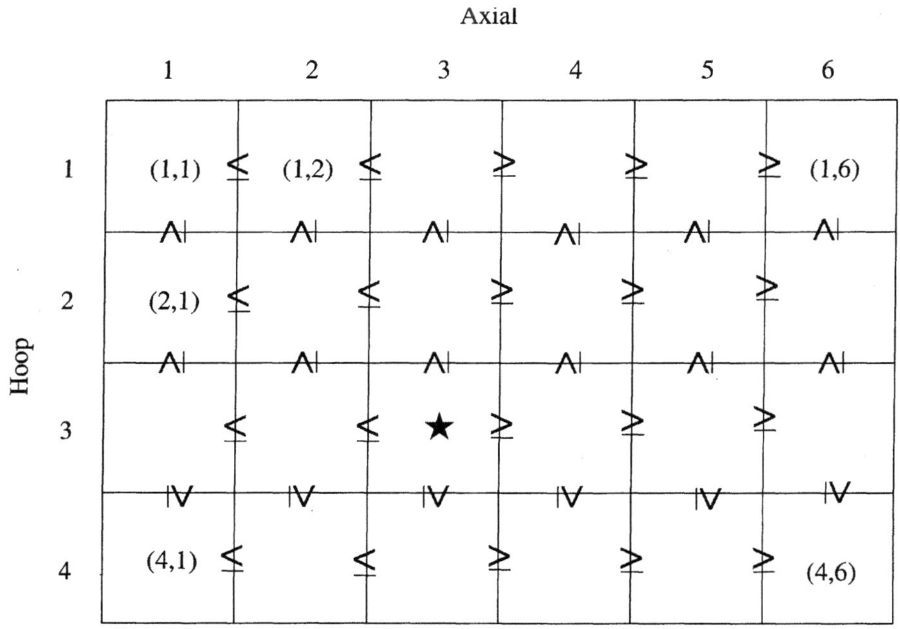

Figure 7.7. The "greater-than-or-equal-to" blending rule applied to a 4 x 6 composite panel.

in later regions. Thus, as one moves further away from the key region, the stacking sequence has fewer and fewer plies. This rule of consistently dropping plies from the key region is called the "greater-than-or-equal-to" blending rule [95, 187]. Figure 7.7 demonstrates how the blending rules are applied to a 4 x 6 panel [95]. The key region (3,3) is marked by a star. The "$\geq$" symbol indicates the direction in which the ply can be dropped. For instance, if a ply exists in region (3,3), it may or may not exist in region (2,3). But if a ply exists in region (2,3), the same ply must exist in region (3,3). Figure 7.8 shows three different ply configurations. Configuration (a) complies with the blending rules, but configurations (b) and (c) do not. In configuration (b), only region (1,4) violates the "greater-than-or-equal-to" blending rules. But regions (1,1), (1,2), and (2,1) in configuration (c) all violate the blending rules.

Extended Formulation: Two different optimization formulations are introduced in [95] to model blending in the optimal design process. The first formulation scheme uses binary thickness variables t to formulate the "greater-than-or-equal-to rule." The second formulation uses integer

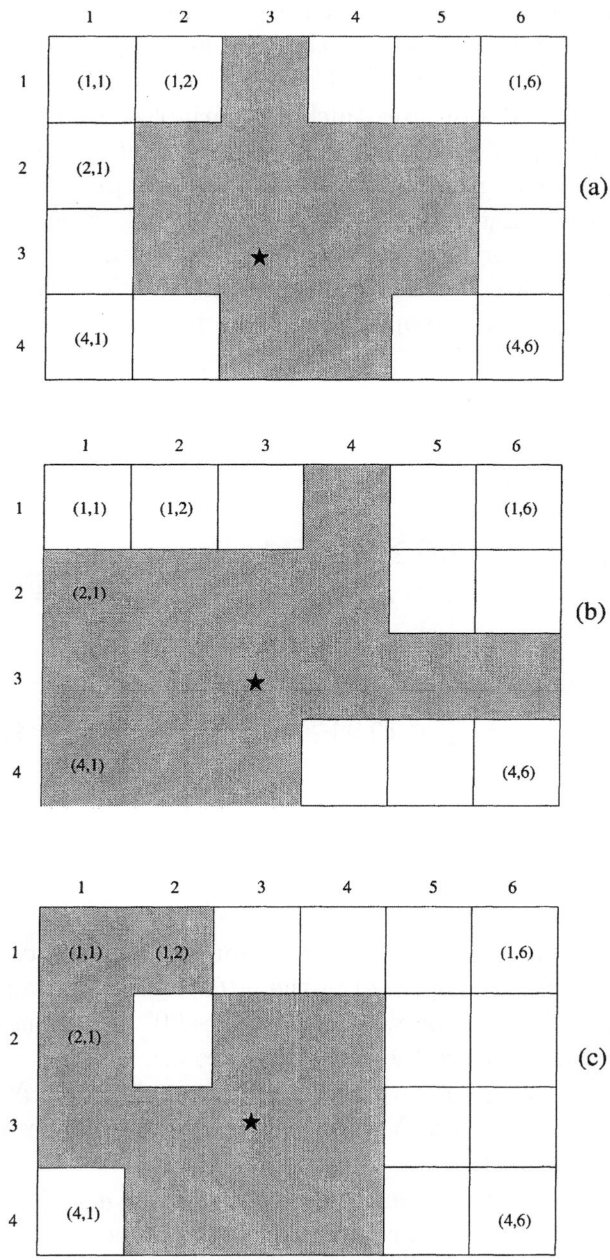

Figure 7.8. Example ply configurations on a 4 x 6 composite panel. Configuration (a) complies with the blending rule but configurations (b) and (c) violate the rule.

distance variables d to denote the number of regions a ply occupies. Consider the $P \times Q$ panel, shown in Figure 7.9, that can have at most N plies. For simplicity assume that the key region is in the upper left corner of the panel, i.e., region (1,1).

The first formulation uses thickness variables. Define a binary thickness variable $t_{i(j,k)}$ such that if $t_{i(j,k)} = 1$ then ply i in region (j,k) exists with fiber angle θ_i, and if $t_{i(j,k)} = 0$, then ply i has been dropped in region (j,k). Let θ represent the vector of θ_i fiber angles and let t represent the vector of $t_{i(j,k)}$ binary thickness variables. The "greater-than-or-equal-to rule" can now be enforced using the thickness variables for a $P \times Q$ panel in the optimization formulation as follows,

$$\begin{array}{lll}
\text{minimize} & f(\theta,t) & \\
\text{subject to} & g_l(\theta,t) \geq 0 & l = 1,\ldots,m \\
& t_{i(j,k)} \geq t_{i(j,k+1)} & i = 1,\ldots,N, \quad j = 1,\ldots,P \\
& t_{i(j,k)} \geq t_{i(j+1,k)} & \quad \text{and } k = 1\ldots,Q \\
& -90 \leq \theta_i \leq 90 & i = 1,\ldots,N \\
& t_{i(j,k)} \in \{0,1\} & i = 1,\ldots,N, \quad j = 1,\ldots,P \\
& & \quad \text{and } k = 1\ldots,Q.
\end{array}$$

In this formulation, the objective function is weight, and the margin of safety constraints are captured in the $g_l(\theta,t) \geq 0$ constraints. The fiber angles are limited to be between $-90°$ and $+90°$. The constraints on the t binary variables ensure the blending rules are satisfied.

The number of design variables in this formulation is $N + (N \cdot P \cdot Q)$ variables, where there are N variables for the fiber angles and $N \cdot P \cdot Q$ binary thickness variables.

In the second formulation, a distance variable $d_{i(j)}$ is defined as the number of regions a ply occupies in a given row j. For example, if $d_{i(j)} = 3$ then ply i extends through the first three regions in row j with fiber angle θ_i. The distance variable $d_{i(j)}$ can take on values between 0 and the number of columns Q. The vector d denotes all $d_{i(j)}$ integer distance variables. Considering a $P \times Q$ panel, there are $N \cdot P$ distance variables, in contrast to $N \cdot P \cdot Q$ thickness variables.

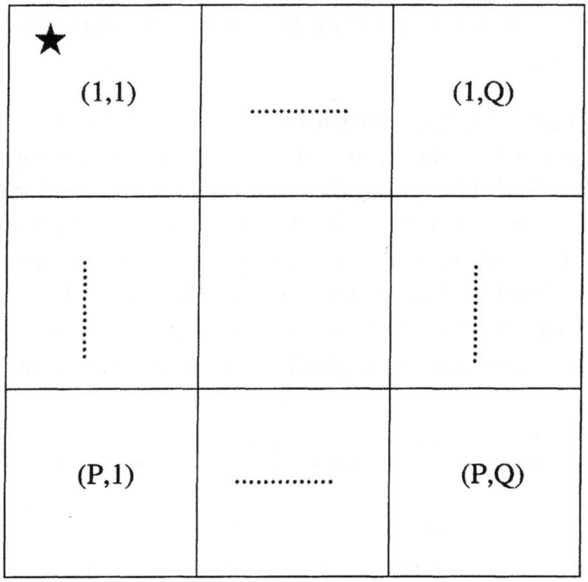

Figure 7.9. Orientation of a $P \times Q$ panel, key region is (1,1)

Using distance variables to enforce the blending rules results in the following formulation,

$$\begin{array}{lll} \text{minimize} & f(\theta, d) & \\ \text{subject to} & g_l(\theta, d) \geq 0 & l = 1, \ldots, m \\ & d_{i(j)} \geq d_{i(j+1)} & i = 1, \ldots, N, \quad j = 1, \ldots, P \\ & d_{i(j)} \in \{0, 1, 2, \ldots, Q\} & i = 1, \ldots, N, \quad j = 1, \ldots, P \\ & -90 \leq \theta_i \leq 90 & i = 1, \ldots, N. \end{array}$$

Again, the objective function is weight, and the margin of safety constraints are captured in the $g_l(\theta, t) \geq 0$ constraints. The fiber angles are constrained to be between $-90°$ and $+90°$. The constraints on the d integer variables also ensure the blending rules are satisfied.

This formulation requires at most $N + (N \cdot P)$ variables, where again there are N variables for the fiber angles, but only $N \cdot P$ distance variables. The distance variables are integer valued, between 0 and Q, whereas the thickness variables are binary valued.

The optimization formulation of a blended panel is a challenging global optimization problem that consists of a mixture of discrete and continuous design variables. This mixed-continuous optimization problem has been solved using Improving Hit-and-Run (IHR) (see Chapter 6, Section 2, or [189]).

Composite panel design problem: To illustrate the usefulness of the blending rules, a (3×4) panel with 12 regions is designed according to the blending rules. The panel dimensions and the load conditions at each region are shown in Figure 7.10. From the figure, region (2,2) is the heaviest loaded region, and is a logical choice for the key region. For more discussion about key elements see [95]. Based on the key element, it is possible to divide the panel into four regions where the "greater-than-or-equal-to" rules can be applied to each one independently.

	96"	92"	105"	105"	
Element	(1,1)	(1,2)	(1,3)	(1,4)	
Nx	3500	3500	3000	3500	
Ny	1000	1000	1000	1500	70.27"
Nxy	300	800	800	1000	
Element	(2,1)	(2,2) ★	(2,3)	(2,4)	
Nx	7000	10000	6000	3000	
Ny	3000	6000	3000	1000	70.27"
Nxy	500	4000	800	500	
Element	(3,1)	(3,2)	(3,3)	(3,4)	
Nx	3500	3000	3000	3500	
Ny	1000	1000	1000	1500	70.27"
Nxy	300	800	800	1000	

Figure 7.10. Loading conditions and panel dimensions for the sample problem.

The design problem considers a sandwich panel that is composed of unidirectional graphite epoxy plies and a honeycomb core, as illustrated in Figure 7.11. The objective is to minimize the weight of the entire panel while maintaining positive margins of safety. The inequality constraints are defined in terms of mechanical performance constraints. Two structural constraints are considered; panel stiffness and panel strength. The panel strength is based on a maximum strain failure criterion which requires that all ply strains must be below specified critical values. The minimum panel stiffness requirements ensure that the panel stiffness

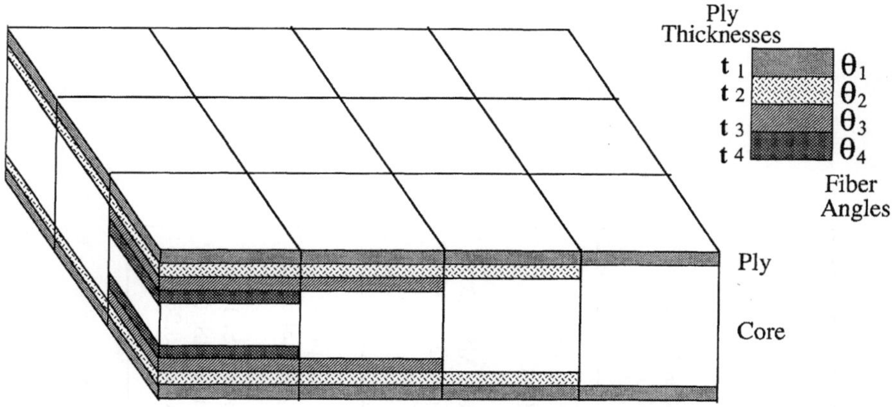

Figure 7.11. Sandwich panel.

must exceed a minimum required stiffness. An explanation of margins of safety computations for these two analysis modules is given in [95].

The sandwich panel has a honeycomb core and the plies in the panel are symmetric about the core. Sometimes it is also required that the facesheet laminates be symmetric about their midplanes. Such a double symmetry implies that the number of plies varies in increments of four. When the blending rules are enforced and the total thickness of the panel stays constant, then the thickness of the core varies between regions. To demonstrate the effectiveness of the blending rules, we first present a design that results when a panel is designed by optimizing each region independently of another region using the IHR algorithm. In this example each region is treated independently and without enforcing any blending rules, hence the name "unblended design."

The unblended design is shown in Figure 7.12. The figure gives the number of plies, their fiber orientation, the most critical strain margin of safety, the most critical stiffness margin of safety, and the core thickness for each region. In this unblended design, the core thickness is held constant at 0.25 in, which implies that the thickness of the entire panel varies as the number of plies changes.

In the blended design shown in Figure 7.13, the core thickness is forced to vary to maintain a constant panel thickness. The fiber angles are allowed to take on integer values. The facesheet plystack is symmetric, and is also symmetric about the core. For instance the ply stack in region $(1,1)$ is given as $[51,-23,21,22]_s$. This means that the plystack $[51,-23,21,22,22,21,-23,51]_s$ is present on both sides of the core.

16 plies $[51, -23, 21, 22]_s$ Strain MS: 3.324 Stiffness MS: 0.009 Core: 0.250	16 plies $[-24, 42, 13, 27]_s$ Strain MS: 3.277 Stiffness MS: 0.023 Core: 0.250	16 plies $[16, 23, -26, 52]_s$ Strain MS: 3.009 Stiffness MS: 0.030 Core 0.250	16 plies $[63, -25, 30, 16]_s$ Strain MS: 1.869 Stiffness MS: 0.0007 Core 0.250
16 plies $[-39, 54, -6, 20]_s$ Strain MS: 0.222 Stiffness MS: 0.034 Core: 0.250	24 plies ★ $[56, 49, -8, 27, -9, 61]_s$ Strain MS: 0.709 Stiffness MS: 0.016 Core: 0.250	20 plies $[58, -24, -24, 58, 12]_s$ Strain MS: 0.983 Stiffness MS: 0.086 Core: 0.250	16 plies $[-2, 48, -27, 32]_s$ Strain MS: 3.318 Stiffness MS: 0.015 Core: 0.250
16 plies $[-6, 46, 20, -34]_s$ Strain MS: 2.610 Stiffness MS: 0.081 Core: 0.250	16 plies $[16, 23, -26, 52]_s$ Strain MS: 3.009 Stiffness MS: 0.030 Core: 0.250	16 plies $[16, 23, -26, 52]_s$ Strain MS: 3.009 Stiffness MS: 0.030 Core: 0.250	16 plies $[63, -25, 30, 16]_s$ Strain MS: 1.869 Stiffness MS: 0.0007 Core: 0.250

Figure 7.12. Layup for an unblended panel: weight is 643.0 lbs.

20 plies $[86, 14, 52, 43, -20]_s$ Strain MS: 0.667 Stiffness MS: 0.061 Core: 0.367	24 plies $[86, -31, 14, 52, 43, -20]$ Strain MS: 1.202 Stiffness MS: 0.318 Core: 0.338	16 plies $[-31, 14, 43, -20]_s$ Strain MS: 1.277 Stiffness MS: 0.025 Core: 0.396	16 plies $[-31, 14, 43, -20]_s$ Strain MS: 0.448 Stiffness MS: 0.025 Core: 0.396
32 plies $[86, 35, -17, 14, 64, 52, 43, -20]_s$ Strain MS: 0.426 Stiffness MS: 0.728 Core: 0.279	36 plies ★ $[86, 35, -17, -31, 14, 64, 52, 43, -20]_s$ Strain MS: 0.381 Stiffness MS: 0.962 Core: 0.250	24 plies $[35, -31, 14, 64, 43, -20]_s$ Strain MS: 0.733 Stiffness MS: 0.346 Core: 0.338	24 plies $[35, -31, 14, 64, 43, -20]$ Strain MS: 2.913 Stiffness MS: 0.346 Core: 0.338
20 plies $[86, 35, -17, 64, -20]_s$ Strain MS: 0.718 Stiffness MS: 0.071 Core: 0.367	28 plies $[86, 35, -17, 14, 64, 43, -20]_s$ Strain MS: 2.428 Stiffness MS: 0.673 Core: 0.308	20 plies $[35, 14, 64, 43, -20]_s$ Strain MS: 2.329 Stiffness MS: 0.105 Core: 0.367	20 plies $[35, 14, 64, 43, -20]_s$ Strain MS: 1.699 Stiffness MS: 0.109 Core: 0.367

Figure 7.13. Layup for a blended panel using t variables: weight is 882.4 lbs.

For any specific region, the strain margin of safety reported in Figures 7.13 and 7.12 is the smallest margin of safety value associated with any strain constraint for any ply under any load condition in that region. Similarly, the stiffness margin of safety in the figures is the most critical margin of safety associated with panel stiffness in that region, i.e., the smallest margin of safety value associated with stiffness in the axial, hoop and shear directions. In the unblended design, the stiffness margin of safety is more critical than the strain margin of safety in each region.

The unblended design shown in Figure 7.12 has comparatively low weight (643 lbs) but cannot be manufactured. The unblended panel considers no coupling between regions during the analysis, so the number of plies, ply fiber angles, and total panel thickness all vary abruptly from one region to the next, to account for load variation from one region to the next. For example, for the loads present in region (1,1) the best design found involves two 8-ply $[51, -23, 21, 22]_s$ laminates, whereas for the immediately adjacent (1,2) region the design involves two 8-ply $[-24, 42, 13, 27]_s$ laminates. In practice, this panel cannot be manufactured due to these abrupt changes in all ply fiber angles. Also, because a constant core thickness of 0.250 in has been assumed, the total panel thickness varies as the number of plies present changes from one region to the next. This variation in panel thickness is also objectionable in some cases. These shortcomings in panel design can be avoided through implementation of the blending rules, as illustrated next.

The Improving Hit-and-Run optimization algorithm was used to design the unblended and blended panels. A fixed number of 5000 iterations were performed using both the hyperspherical direction generator (HD) and coordinate direction generator (CD) (see Chapter 6, Section 4.1). The two different formulations were compared in [95]; one formulation using the binary thickness variables t, and the second using the integer distance variables d.

Figure 7.13 shows the final design derived from the t formulation [95]. For this panel, the changes from region to region are more gradual than the unblended design, reducing the abruptness of the design and increasing the validity of the analysis performed. Also, with common fiber angles appearing in adjacent regions the panel is manufacturable. Region (2,2) is the key region and has 36 plies which is the highest number of plies in the panel. In fact, all the fiber angles in other regions are a subset of the fiber angles for those in the key region. Each ply has the same fiber angle throughout the panel and satisfies the blending rules. For instance, a -20 degree ply exists in all the regions and a 64 degree ply exists in regions (2,1) through (3,4). Each ply can be easily cut out or laid down and the panel consists of stacked up connected plies.

The weight of the blended panel (882.4 lbs) is significantly higher than that of the unblended panel (643.0 lbs). This apparent increase in weight is misleading since in practice the unblended panel cannot be manufactured. The unblended analysis is still useful because it represents a lower bound on the panel weight.

In [95], the d formulation is also applied to find a design for a blended panel. In fact, the d formulation finds a lower weight panel than the t formulation. The d formulation requires fewer design variables and therefore it is easier for the optimizer to find a better design within the fixed 5,000 iterations. The t formulation uses 14 variables for fiber angles, which allows up to 28 plies on each side of the core, and maximum of 56 plies total due to symmetry. Hence there are $14 \cdot 12 = 168$ binary thickness variables for the blended panel. In contrast, the d formulation also has 14 variables for fiber angles, but only 9 integer distance variables per ply, totaling $14 \cdot 9 = 126$ integer distance variables.

Typically fewer variables indicate that the computation time for the d formulation is shorter than for the t formulation. If we use the same amount of computation, the d formulation achieves a better design (lower weight in this case) than the t formulation. The blended designs are difficult to obtain without the use of an automated optimization tool.

In summary, this example demonstrates the benefit of the blending rules as opposed to a point optimized design in the unblended panel. In [95], a larger example is presented where the difference between the t and d formulations are especially pronounced. The t formulation requires 5100 design variables while the d formulation requires only 300 design variables. The difference in number of design variables has a profound impact on computation time, with the d formulation finding a solution much faster.

5. Summary

This chapter demonstrates how a stochastic optimization algorithm can be utilized in practical global optimization problems. Engineering examples, such as the fuel allocation problem and the truss design problem often include constraints. The hierarchical formulation and the penalty function approach are discussed to handle nonlinear inequality constraints.

Experience with optimizing the design of composite structures is described. The design of composite structures is formulated as a mixed continuous/integer global optimization problem including manufacturing constraints. The optimization approach has been demonstrated in practice to obtain solutions that were difficult to find in the past. Moreover, the optimization approach implemented is efficient in the sense that it found solutions, albeit sub-optimal ones, within a reasonable amount of time, and offers design engineers a systematic way to evaluate various designs. The blending rules force the design to be practical from a manufacturing point of view when a large panel experiences a range of loads. The incorporation of both discrete and continuous variables allows integer or binary variables to capture the number of plies, and is also useful in exploring the impact of discretizing variables such as fiber angles. In sum, these methods create a wealth of opportunities to optimize practical problems that were previously beyond reach.

References

[1] Aarts, E., and Korst, J., *Simulated Annealing and Boltzmann Machines: A Stochastic Approach to Combinatorial Optimization and Neural Computing,* John Wiley and Sons, Chichester, England, 1989.

[2] Aldous, D.J., and Fill, J., *Reversible Markov Chains and Random Walks on Graphs,* monograph in preparation, http://www.stat.berkeley.edu/users/aldous/book.html, 2003.

[3] Ali, M.M., Khompatraporn, C., and Zabinsky, Z.B., "A Numerical Evaluation of Several Global Optimization Algorithms on Selected Benchmark Test Problems," submitted for publication, 2003.

[4] Ali, M.M., Törn, A., "Population Set Based Global Optimization Algorithms: Some Modifications and Numerical Studies," submitted to *Computers and Operations Research,* 2002.

[5] Abramowitz, M., and Stegun, I.A., editors, *Handbook of Mathematical Functions with Formulas, Graphs, and Mathematical Tables,* National Bureau of Standards, Applied Mathematics Series 55, June 1964.

[6] Anderssen, R.S., "Global Optimization," in *Optimization,* edited by R.S. Anderssen, L.S. Jennings and D.M. Ryan, University of Queensland Press, 1-15, 1972.

[7] Archetti, F., Betro, B., and Steffe, S., "A Theoretical Framework for Global Optimization Via Random Sampling," Working Paper, Quaderno dei gruppi di recerca matematica del C.N.R., University of Pisa, 1975.

[8] Arora, J.S., "Computational Design Optimization: A Review and Future Directions," *Structural Safety 7,* 131-148, 1990.

[9] Arora, J.S., editor, *Guide to Structural Optimization,* ASCE Manuals and Reports on Engineering Practice No. 90, American Society of Civil Engineers, New York, 1997.

[10] Baier, H.J., "Composite Laminate and Sandwich Optimization with Application," in *Optimization of Large Structural Systems, Volume II,* edited by G.I.N. Rozvany, Kluwer Academic Publishers, Netherlands, 997-1009, 1993.

[11] Baritompa, W.P., Baoping, Z., Mladineo, R.H., Wood, G.R., and Zabinsky, Z.B., "Towards Pure Adaptive Search," *Journal of Global Optimization 7,* 93-110, 1995.

[12] Bartle, R.G., and Sherbert, D.R., *Introduction to Real Analysis,* John Wiley and Sons, New York, 1982.

[13] Bazaraa, M.S., Sherali, H.D., and Shetty, C.M., *Nonlinear Programming Theory and Algorithms,* John Wiley and Sons, New York, 1993.

[14] Bélisle, C.J.P., "Convergence Theorems for a Class of Simulated Annealing Algorithms on $\Re^d$," *Journal of Applied Probability 29,* 885-895, 1992.

[15] Bélisle, C.J.P., "Slow Hit-and-Run Sampling," *Statistics & Probability Letters 47,* 33-43, 2000.

[16] Bélisle, C.J.P., Romeijn, H.E., and Smith, R.L., "Hit-and-Run Algorithms for Generating Multivariate Distributions," *Mathematics of Operations Research 18,* 255-266, 1993.

[17] Belton, B. editor, "Math That Helps Nearly Everyone Make Decisions," *Business Week Online*, at website http://www.businessweek.com/bwdaily/dnflash/oct2000/nf20001013_580.htm, October 13, 2000.

[18] Bennage, W.A., and Dhingra, A.K., "Single and Multiobjective Structural Optimization in Discrete-Continuous Variables using Simulated Annealing," *International Journal for Numerical Methods in Engineering 38,* 2753-2773, 1995.

[19] Berbee, H.C.P., Boender, C.G.E. , Rinnooy Kan, A.H.G., Scheffer, C.L., Smith, R.L., and Telgen, J., "Hit-and-Run Algorithms for the Identification of Nonredundant Linear Inequalities," *Mathematical Programming 37,* 184-207, 1987.

[20] Berthelot, J-M., *Composite Materials: Mechanical Behavior and Structural Analysis*, Springer-Verlag, Inc., New York, 1999.

[21] Bhaskar, K., and Librescu, L., "Buckling Under Axial Compression of Thin-Walled Composite Beams Exhibiting Extension-Twist Coupling," *Composite Structures 31,* 203-212, 1995.

[22] Birge, J.R., and Smith, R.L., "Random Procedures for Nonredundant Constraint Identification in Stochastic Linear Programs," *Mathematical and Management Sciences 4,* 41-70, 1984.

[23] Boender, C.G.E., and Romeijn, H.E., "Stochastic Methods," in *Handbook of Global Optimization*, edited by R. Horst, and P.M. Pardalos, Kluwer Academic Publishers, Netherlands, 829-869, 1995.

[24] Boneh, A., "A Probabilistic Algorithm for Identifying Redundancy by a Random Feasible Point Generator (RFPG)," in *Redundancy in Mathematical Programming,* edited by M.H. Karwan, B. Lotfi, J. Telgen and S. Zionts, Springer-Verlag, Berlin, 1983.

[25] Boneh, A., and Golan, A., "Constraints' Redundancy and Feasible Region Boundedness by Random Feasible Point Generator," *Third European Congress on Operations Research, EURO III*, Amsterdam, April 9-11, 1979.

[26] Brooks, S.H., "A Discussion of Random Methods for Seeking Maxima," *Operations Research 6*, 244-251, 1958.

[27] Bulger, D.W., Alexander, D., Baritompa, W.P., Wood, G.R., and Zabinsky, Z.B., "Expected Hitting Times for Backtracking Adaptive Search," Technical Report, IIST, Masey University, Palmerston North, New Zealand, 2001.

[28] Bulger, D.W., Baritompa, W.P., and Wood, G.R, "Implementing Pure Adaptive Search with Grover's Quantum Algorithm," *Journal of Optimization Theory and Applications, 116*, 517-529, 2003.

[29] Bulger, D.W., and Wood, G.R., "Hesitant Adaptive Search for Global Optimisation," *Mathematical Programming 81*, 89-102, 1998.

[30] Bulger, D.W., and Wood, G.R., "On the Convergence of Localisation Search," in *State of the Art in Global Optimization: Computational Methods and Applications*, edited by C.A. Floudas and P.M. Pardalos, Kluwer Academic Publishers, 227-233, 1996.

[31] Cardosa, M.F., Salcedo, R.L., and Azevedo, S.F., "Simplex-Simulated Annealing Approach to Continuous Non-linear Optimization," *Computers & Chemical Engineering 20 (9),* 1065-1080, 1996.

[32] Caron, R.J., Hlynka, M., and McDonald, J.F., "On the Best Case Performance of Hit and Run Methods for Detecting Necessary Constraints," *Mathematical Programming 54,* 233-249, 1992.

[33] Chandra, R., and Chopra, I., "Experimental and Theoretical Analysis of Composite I-Beams with Elastic Couplings," *AIAA Journal 29 (12),* 1991.

[34] Chen, M.-H, and Schmeiser, B.W., "General Hit-and-Run Monte Carlo Sampling for Evaluating Multidimensional Integrals," *Operations Research Letters 19,* 161-169, 1996.

[35] Clough, D.J., "An Asymptotic Extreme-Value Sampling Theory for Estimation of a Global Maximum," *CORS Journal 7*, 102-115, 1969.

[36] Corana, A., Marchesi, M., Martini, C., and Ridella, S., "Minimizing Multimodal Function of Continuous Variables with the Simulated Annealing Algorithm," *ACM Transactions on Mathematical Software 13 (3),* 262-280, 1987.

[37] Craddock, J. N. and Yen, S. C., "The Bending Stiffness of Laminated Composite Material I-Beams," *Composites Engineering,* Vol. 3, No. 11, 1993.

[38] Cramér, H., *Mathematical Methods of Statistics*, Princeton University Press, 1946.

[39] De Haan, L., "Estimation of the Minimum of a Function Using Order Statistics," *Journal of the American Statistical Association 76*, 467-469, 1981.

[40] De Jong, K.A., Spears, W.M., and Gordon, D.F., "Using Markov Chains to Analyze GAFOs," in *Foundations of Genetic Algorithms 3,* edited by D. Whitley, and M. Vose, Morgan Kaufmann, San Francisco, 115-137, 1995.

[41] Dekkers, A., and Aarts, E., "Global Optimization and Simulated Annealing," *Mathematical Programming 50,* 367-393, 1991.

[42] Dereksdottir, E.H., "A Computational Study of Improving Hit-and-Run Algorithms on a Discrete Domain," *Master's Thesis,* Industrial Engineering, University of Washington, 2002.

[43] Devore, J.L., *Probability and Statistics for Engineering and the Sciences, Second Edition,* Wadsworth Inc, California, 1987.

[44] Dixon, L.C.W., and Szegö, G.P., editors, *Towards Global Optimization,* North-Holland, Amsterdam, 1975.

[45] Dixon, L.C.W., and Szegö, G.P., editors, *Towards Global Optimization 2,* North-Holland, Amsterdam, 1978.

[46] Dowsland, K.A., "Simulated Annealing," in *Modern Heuristic Techniques for Combinatorial Problems,* edited by C.R. Reeves, Halsted Press, New York, 20-69, 1993.

[47] Dueck, D., and Scheuer, T., "Threshold Accepting: a general purpose optimization algorithm appearing superior to simulated annealing," *J. Comp. Physics 90,* 161-175, 1990.

[48] Dyer, M.E., and Frieze, A.M., "Computing the Volume of Convex Bodies: A Case Where Randomness Provably Helps," *Proceedings of Symposia in Applied Mathematics 44,* 123-169, 1991.

[49] Feller, W., *An Introduction To Probability Theory And Its Applications, Volume 2, Second Edition,* John Wiley and Sons, New York, 1971.

[50] Floudas, C.A., and Pardalos, P.M., editors, *Recent Advances in Global Optimization,* Princeton University Press, 1992.

[51] Floudas, C.A., and Visweswaran, V., "A Global Optimization Algorithm (GOP) for Certain Classes of Nonconvex NLPs - I. Theory," *Computers and Chemical Engineering 14,* 1397, 1990.

[52] Freund, R.M., "Polynomial-time Algorithms for Linear Programming Based Only on Primal Scaling and Projected Gradients of a Potential Function," MIT working paper OR-182-88, Massachusetts Institute Of Technology, Cambridge, 1988.

[53] Galambos, J., *The Asymptotic Theory Of Extreme Order Statistics,* John Wiley and Sons, New York, 1978.

[54] Garey, M.R., and Johnson, D.S., *Computers and Intractability: A Guide to the Theory of NP-completeness,* WH Freeman and Co., San Francisco, 1979.

[55] Glover, F., and Laguna, M., "Tabu Search," in *Modern Heuristic Techniques for Combinatorial Problems,* edited by C.R. Reeves, Halsted Press, New York, 70-150, 1993.

[56] Gonzaga, C.C., "Polynomial Affine Algorithms for Linear Programming," Report ES-139/88, Federal University of Rio de Janeiro, Rio de Janeiro, 1988.

[57] Gradshteyn, I.S., and Ryzhik, I.M., editors, translated by Alan Jeffrey, *Table of Integrals, Series, and Products*, Academic Press, New York, 1980.

[58] Graesser, D.L., Zabinsky, Z.B., Tuttle, M.E., and Kim, G.I., "Designing Laminated Composites Using Random Search Techniques," *Composite Structures 18*, 311-325, 1991.

[59] Graesser, D.L., Zabinsky, Z.B., Tuttle, M.E., and Kim, G.I., "Optimal Design of Composite Structures," *Composite Structures 24*, 273-281, 1993.

[60] Gürdal, Z., Haftka, R.T., "Optimization of Composite Laminates," in *Optimization of Large Structural Systems, Volume II*, edited by G.I.N. Rozvany, Kluwer Academic Publishers, Netherlands, 623-648, 1993.

[61] Gürdal, Z., Haftka, R.T., Hajela, P., *Design and Optimization of Laminated Composite Materials*, John Wiley & Sons, Inc., New York, 102-221, 1999.

[62] Haftka, R.T., and Gürdal, Z., *Elements of Structural Optimization*, Third Revised and Expanded Edition, Kluwer Academic Publishers, Netherlands, 1992.

[63] Halton, J.H., "On the Efficiency of Certain Quasi-random Sequences of Points in Evaluating Multi-dimensional Integrals," *Numerische Mathematik 2,* 84-90, 1960.

[64] Hansen, P., Jaumard, B., and Lu, S-H., "Local Optimization of Univariate Lipschitz Functions: I. Survey and Properties," *Mathematical Programming 55,* 251-272, 1992.

[65] Harris, B., *Theory of Probability*, Addison-Wesley, Massachusetts, 1966.

[66] Haug, E.J., and Arora, J.S., *Applied Optimal Design - Mechanical and Structural Systems*, John Wiley and Sons, New York, 1979.

[67] Heglund, D.A., "Methodology for Efficient Random Search in the Optimization of Composite Laminate Designs," *Master's Thesis*, University of Washington, 1992.

[68] Heglund, D.A., Zabinsky, Z.B., Tuttle, M.E., Graesser, D.L., and Kim, G.I., "Optimization in Preliminary Structural Design," *Proceedings of the 15th Annual IIE Aerospace and Defense Conference*, 1992.

[69] Hendrix, E.M.T., and Klepper, O., "On Uniform Covering, Adaptive Random Search and Raspberries," *Journal of Global Optimization 18 (2)*, 143-163, 2000.

[70] Hibbeler, R.C., *Structural Analysis, Third Edition*, Prentice Hall, New Jersey, 1997.

[71] Holland, J., Workshop on Operations Research and Complex Adaptive Systems, Santa Fe Institute, Santa Fe, New Mexico, May 1992.

[72] Horst, R., and Pardalos, P.M., editors, *Handbook of Global Optimization,* Kluwer Academic Publishers, Netherlands, 1995.

[73] Horst, R., and Tuy, H., *Global Optimization: Deterministic Approaches*, Second Edition, Springer, Berlin, 1993.

[74] Hovanessian, S.A., and Pipes, L.A., *Digital Computer Methods in Engineering,* McGraw Hill, New York, 1969.

[75] Hovanessian, S.A., and Stout, T.M., "Optimum Fuel Allocation in Power Plants," *IEEE Transactions on Apparatus Systems 66*, 329-335, 1963.

[76] Hsieh, Y-Y., *Elementary Theory of Structures, Third Edition,* Prentice Hall, New Jersey, 1988.

[77] Huard, P., "Resolution of Mathematical Programming with Non-Linear Constraints by the Method of Centers," in *Non-Linear Programming,* edited by J. Abadie, North-Holland, Amsterdam, 207-219, 1967.

[78] Jones, R.M., *Mechanics of Composite Materials,* Scripta Book Company, Washington D.C, 1975.

[79] Karmarkar, N., "A New Polynomial-time Algorithm for Linear Programming," *Combinatorica 4*, 373-395, 1984.

[80] Karwan, M.H., Lotfi, V., Telgen, J., and Zionts, S., editors, *Redundancy in Mathematical Programming,* Springer-Verlag, Berlin, 108-134, 1983.

[81] Kaufman, D.E., and Smith, R.L., "Direction Choice for Accelerated Convergence in Hit-and-Run Sampling," *Operations Research 46,* 84-95, 1998.

[82] Kemeny, J.G., and Snell, J.L., *Finite Markov Chains,* Springer-Verlag, New York, 1976.

[83] Kendall, M.G., *A Course In The Geometry Of n-Dimensions*, Hafner Publishing, New York, 1961.

[84] Khompatraporn, C., Pintér, J.D., and Zabinsky, Z.B., "Comparative Assessment of Algorithms and Software for Global Optimization," submitted for publication, 2002.

[85] Kiatsupaibul, S., "Markov Chain Monte Carlo Methods for Global Optimization," *Dissertation*, University of Michigan, 2000.

[86] Kiatsupaibul, S. and Smith, R.L., "An Analytically Derived Cooling Schedule for Simulated Annealing," Technical Report, Department of Industrial and Operations Engineering, The University of Michigan, Ann Arbor, Michigan 48109, June 26, 2000.

[87] Kiatsupaibul, S., Smith, R.L., and Zabinsky, Z.B., "A Discrete Hit-and-Run Algorithm for Generating Samples from General Discrete Multivariate Distributions," Technical Report, Department of Industrial and Operations Engineering, The University of Michigan, Ann Arbor, Michigan 48109, 2001.

[88] Kingman, J.F.C., *Poisson Processes*, Oxford University Press, Oxford, England, 1993.

[89] Kirkpatrick, S., Gelatt Jr., C.D., and Vecchi, M.P., "Optimization by Simulated Annealing," *Science 220*, 671-680, 1983.

[90] Knuth, D.E., *The Art of Computer Programming, Volume 2,* Addison-Wesley, Reading, Massachusetts, 116, 1969.

[91] Koehler, G.J., "New Directions in Genetic Algorithm Theory," *Annals of Operations Research 75,* 49-68, 1997.

[92] Koehler, G.J., "Computing Simple GA Expected Waiting Times," *Proceedings of the Genetic and Evolutionary Computation Conference,* 795, 1999.

[93] Kristinsdottir, B.P., "Complexity Analysis of Random Search Algorithms," *Dissertation*, University of Washington, 1997.

[94] Kristinsdottir, B.P., Zabinsky, Z.B., Tuttle, M.E., and Csendes, T., "Incorporating Manufacturing Tolerances in Near-Optimal Design of Composite Structures," *Engineering Optimization 26*, 1-23, 1996.

[95] Kristinsdottir, B.P., Zabinsky, Z.B., Tuttle, M.E., and Neogi, S., "Optimal Design of Large Composite Panels with Varying Loads," *Composite Structures 51,* 93-102, 2001.

[96] Kristinsdottir, B.P., Zabinsky, Z.B., and Wood, G.R., "Discrete Backtracking Adaptive Search for Global Optimization," a book chapter in *Stochastic and Global Optimization*, edited by G. Dzemyda, V. Šaltenis, and A. Žilinskas, Kluwer Academic Publishers, 147-174, 2002.

[97] Lawrence, J.P. III, and Steiglitz, K., "Randomized pattern search," *IEEE Transactions On Computers C-21*, 382-385, 1972.

[98] Lee, L.H., Poolla, K., "On Statistical Model Validation," *Transactions of the ASME 118*, 226-236, 1996.

[99] Locatelli, M., "Convergence Properties of Simulated Annealing for Continuous Global Optimization," *Journal of Applied Probability 33*, 1127-1140, 1996.

[100] Locatelli, M., "Convergence and First Hitting Time of Simulated Annealing Algorithms for Continuous Global Optimization," *Math Method Operations Research 54*, 171-199, 2001.

[101] Locatelli, M., and Schoen, F., "Simple Linkage: Analysis of a Threshold-accepting Global Optimization Method," *Journal of Global Optimization 9*, 95-111, 1996.

[102] Lovász, L., "Hit-and-Run Mixes Fast," *Mathematical Programming 86,* 443-461, 1999.

[103] Luenberger, D.G., *Linear and Nonlinear Programming*, Second Edition, Addison-Wesley, Massachusetts, 1984.

[104] Luus, R., and Jaakola, T.H.I., "Optimization by Direct Search and Systematic Reduction of the Size of Search Region," *A.I.Ch.E. Journal 19(4)*, 760-766, 1973.

[105] Mabson, G.E., Flynn, B.W., Ilcewicz, L.B., and Graesser, D.L., "The Use of COSTADE in Developing Composite Commercial Aircraft Fuselage Structures," *Proceedings of the 35th AIAA/ASME/ASCE/AHS/ASC Structure, Structural Dynamics, and Materials Conference*, April 1994.

[106] Mabson, G.E., Fredrikson, H.G., Graesser, D.L., Metschan, S.L., Proctor, M.R., Stogin, D.C., Tervo, D.K., Tuttle, M.E., Zabinsky, Z.B., and Gutowski, T.G., "Cost Optimization Software for Transport Aircraft Design Evaluation," presented at the *Sixth NASA/DOD/ARPA Advanced Composite Technology Conference*, 1995.

[107] Mabson, G.E., and Graesser, D.L., *Cost Optimization Software for Transport Aircraft Design Evaluation (COSTADE) - Users Manual*, NASA Contractor Report 4738, August 1996.

[108] Manchanda, P., Ansari, A., and Gupta, S., "The 'Shopping Basket': A Model for Multicategory Purchase Incidence Decisions," *Marketing Science 18*, 95-114, 1999.

[109] Massard, T.N., "Computer Sizing of Composite Laminates for Strength," *J. Reinforced Plastics Comp. 3*, 300-345, 1984.

[110] Metropolis, N., Rosenbluth, A.W., Rosenbluth, M.N., Teller, A.H., and Teller, E., "Equations of State Calculations by Fast Computing Machines," *The Journal of Chemical Physics 21*, 1087–1092, 1953.

[111] Metscham, S., Mabson, G.E., and Graesser, D.L., "Designing a Window Belt," *Proceedings of the 5th NASA Advanced Composite Technology Conference*, August 1994.

[112] Morton, S.K., and Webber, J.P.H., "Optimal Design of a Composite I-Beam," *Composite Structures 28,* 149-168, 1994.

[113] Murty, K.G., *Linear Programming*, John Wiley and Sons, New York, 1983.

[114] Mutseniyeks, V.A., and Rastrigin, L., "Extremal Control of Continuous Multiparameter Systems by the Method of Random Search," *Engineering Cybernetics 1*, 82-90, 1964.

[115] Nagendra, S., Jestin, D., Gürdal, Z., Haftka, R.T., and Watson, L.T., "Improved Genetic Algorithm for the Design of Stiffened Composite Panels," *Proceedings of the 15th International Symposium on Mathematical Programming*, 1994.

[116] Neogi, S., "Design of Large Composite Structures Using Global Optimization and Finite Element Analysis," *Dissertation*, University of Washington, 1997.

[117] Neogi, S., Zabinsky, Z.B., and Tuttle, M.E., "Optimal Design of Composites Using Mixed Discrete and Continuous Variables," *Proceedings of the ASME*

Winter Annual Meeting, Symposium on Processing, Design and Performance of Composite Materials, MD-52, 91-107, 1994.

[118] Neogi, S., Zabinsky, Z.B., Tuttle, M.E., and Kristinsdottir, B.P., "Load Redistribution Issues in Optimal Design of Composites Airplane Structures," *Proceedings of the ICCE/2 Conference*, New Orleans, 1995.

[119] NEOS Guide Optimization Tree, http://www-fp.mcs.anl.gov/otc/Guide /OptWeb/index.html, February 2003.

[120] Newton, M.A., Guttorp, P., Sandra, C., Assunção, R., Abkowitz, J.L., "Stochastic Modeling of Early Hematopoiesis," *Journal of American Statistical Association, 90*, 1146-1155, 1995.

[121] Orlin, J.,"Very Large Neighborhood Search," Tutorial presented at INFORMS San Jose, November 2002.

[122] Pardalos, P.M., and Romeijn, H.E., editors, *Handbook of Global Optimization, Volume 2,* Kluwer Academic Publishers, Netherlands, 2002.

[123] Park, W.J., "An Optimal Design of Simple Symmetric Laminates under the First Ply Failure Criteria," *J. Comp. Mater. 16*, 341-355, 1982.

[124] Patel, N.R., and Smith, R.L., "The Asymptotic Extreme Value Distribution of the Sample Minimum of a Concave Function under Linear Constraints", *Operations Research 31*, 789-794, 1983.

[125] Patel, N.R., Smith, R.L., and Zabinsky, Z.B., "Pure Adaptive Search in Monte Carlo Optimization," *Mathematical Programming 4*, 317-328, 1988.

[126] Pinedo, M., *Scheduling - Theory, Algorithms, and Systems*, Prentice Hall, New Jersey, 1995.

[127] Pintér, J.D., *Global Optimization in Action, Continuous and Lipschitz Optimization: Algorithms, Implementations and Applications*, Kluwer Academic Publishers, Netherlands, 1996.

[128] Pintér, J.D.,"Global Optimization: Software, Test Problems, and Applications," in *Handbook of Global Optimization, Volume 2,* edited by P.M. Pardalos, and H.E. Romeijn, Kluwer Academic Publishers, Netherlands, 231-249, 2002.

[129] Rardin, R.L., *Optimization in Operations Research*, Prentice Hall, New Jersey, 1998.

[130] Ravindran, A., Phillips, D.T., and Solberg, J.J., *Operations Research, Principles and Practice*, John Wiley and Sons, 1976.

[131] Rastrigin, L.A., "Extremal Control by the Method of Random Scanning," *Automation and Remote Control 21*, 891-896, 1960.

[132] Rastrigin, L.A., "The Convergence of the Random Method in the Extremal Control of a Many-parameter System," *Automation and Remote Control 24*, 1337-1342, 1963.

[133] Reaume, D.J., Romeijn, H.E., and Smith, R.L., "Implementing Pure Adaptive Search for Global Optimization Using Markov Chain Sampling," *Journal of Global Optimization 20,* 33-47, 2001.

[134] Renegar, J., "A Polynomial-time Algorithm, Based on Newton's Method, for Linear Programming," *Mathematical Programming 40*, 59-93, 1988.

[135] Riche, R.L., and Haftka, R.T., "Optimization of Laminated Stacking Sequence for Buckling Load Maximization by Genetic Algorithm," *AIAA Journal 31 (5),* 951-956, May 1993.

[136] Rinnooy Kan, A.H.G., and Timmer, G.T., "Stochastic Methods for Global Optimization," *American Journal of Mathematical and Management Sciences 4*, 7-40, 1984.

[137] Rinnooy Kan, A.H.G., and Timmer, G.T., "Stochastic Global Optimization Methods Part I: Clustering Methods," *Mathematical Programming 39*, 27-56, 1987.

[138] Rinnooy Kan, A.H.G., and Timmer, G.T., "Stochastic Global Optimization Methods Part II: Multi-level Methods," *Mathematical Programming 39*, 57-78, 1987.

[139] Romeijn, H.E., and Smith, R.L., "Simulated Annealing and Adaptive Search in Global Optimization," *Probability in the Engineering and Informational Sciences 8,* 571-590, 1994.

[140] Romeijn, H.E., and Smith, R.L., "Simulated Annealing for Constrained Global Optimization," *Journal of Global Optimization 5,* 101-126, 1994.

[141] Romeijn, H.E., Zabinsky, Z.B., Graesser, D.L., and Neogi, S., "New Reflection Generator for Simulated Annealing in Mixed-Integer/Continuous Global Optimization," *Journal of Optimization: Theory and Applications 101 (2),* 403-427, 1999.

[142] Ross, S.M., *Stochastic Processes*, John Wiley and Sons, New York, 1983.

[143] Rubinstein, R.Y., "Generating Random Vectors Uniformly Distributed Inside and on the Surface of Different Regions," *European Journal of Operations Research 10*, 205-209, 1982.

[144] Rudolph, G., "Convergence Analysis of Canonical Genetic Algorithms," *IEEE Transactions on Neural Networks 5*, 96-101, 1994.

[145] Savic, V., Tuttle, M.E., and Zabinsky, Z.B., "Optimization of Composite I-Sections Using Fiber Angles as Design Variables," *Composite Structures 53,* 265-277, 2001.

[146] Scales, L.E., *Introduction to Non-linear Optimization*, Macmillan, 1985.

[147] Schmit, L.A., "Structural Synthesis—Its Genesis and Development," *AIAA J. 19*, 1249-1263, 1981.

[148] Schmit, L.A., and Farshi, B., "Optimum Laminate Design for Strength and Stiffness," *Int. J. Numer. Meth. Engng. 7*, 519-536, 1973.

[149] Schoen, F., "Two-phase Methods for Global Optimization," in: *Handbook of Global Optimization, Volume 2,* edited by P.M. Pardalos, and H.E. Romeijn, Kluwer Academic Publishers, Netherlands, 151-177, 2002.

[150] Schrack, G., and Borowski, N., "An experimental comparison of three random searches," in *Numerical Methods For Nonlinear Optimization*, edited by F. Lootsma, Academic Press, London, 137-147, 1972.

[151] Schrack, G., and Choit, M., "Optimized Relative Step Size Random Searches," *Mathematical Programming 10*, 270-276, 1976.

[152] Schumer, M.A., and Steiglitz, K., "Adaptive step size random search," *IEEE Transactions On Automatic Control AC-13*, 270-276, 1968.

[153] Seifer, J.D., "Incorporating Realistic Engineering Considerations into the Optimal Design of Composite Structures," *Master's Thesis*, University of Washington, 1995.

[154] Shen, Y., and Zabinsky, Z.B., "Performance Analysis of Annealing Adaptive Search," Technical Report, Industrial Engineering, University of Washington, Seattle, WA, 2003.

[155] Shen, Y., Zabinsky, Z.B., and Smith, R.L., "Discrete Hit-and-Run Algorithms: Theoretical and Computational Studies," INFORMS San Jose, CA, November 2002.

[156] Shi, L., and Ólafsson, S., "Nested Partitions Method for Global Optimization," *Operations Research 48 (3),* 390-407, 2000.

[157] Smith, R.L., "Efficient Monte Carlo Procedures for Generating Points Uniformly Distributed over Bounded Regions," *Operations Research 32,* 1296-1308, 1984.

[158] Smith, S.J., and Bank, L.C., "Modifications to Beam Theory for Bending and Twisting of Open-Section Composite Beams-Experimental Verification," *Composite Structures 22,* 1992.

[159] Solis, F.J., and Wets, R.J.-B., "Minimization by Random Search Techniques," *Mathematics Of Operations Research 6*, 19-30, 1981.

[160] Sommerfeld, A., *Partial Differential Equations in Physics*, Academic Press, New York, 1949.

[161] Stephens, C.P., and Baritompa, W.P., "Global Optimization Requires Global Information," *Journal of Optimization Theory and Applications,* Vol. 96, No. 3, 575-588, 1998.

[162] Stern, J.M., "Simulated Annealing with a Temperature Dependent Penalty Function," *ORSA Journal on Computing*, Vol. 4, No. 3, 311-319, 1992.

[163] Swanson, G.D., Ilcewicz, L.B., Walker, T., Graesser, D.L., Tuttle, M.E., and Zabinsky, Z.B., "Local Design Optimization for Transport Fuselage Crown Panels," *Proceedings of the 9th DoD/NASA/FAA Conference on Fibrous Composites in Structural Design*, Lake Tahoe, Nevada, November 1991.

[164] Thanedar, P.B., Arora, J.S., Li, G.Y., and Lin, T.C., "Robustness, Generality and Efficiency of Optimization Algorithms for Practical Applications," *Structural Optimization 2*, 203-212, 1990.

[165] Törn, A., and Žilinskas, A., *Global Optimization*, Springer-Verlag, Germany, 1989.

[166] Tuttle, M.E., and Zabinsky, Z.B., "Methodologies for Optimal Design of Composite Fuselage Crown Panels," *Proceedings of the 35th AIAA/ASME/ASCE/AHS/ASC Structures, Structural Dynamics, and Material Conference*, April 1994.

[167] Trouvé, A., "Cycle Decompositions and Simulated Annealing," *SIAM J. Control and Optimization 34*, 966-986, 1996.

[168] Vanderbilt, D., and Louie, S.G., "A Monte Carlo Simulated Annealing Approach to Optimization over Continuous Variables," *Journal of Computational Physics 56*, 259-271, 1984.

[169] Vanderplaats, G.N., "Structural Optimization—Past, Present, and Future," *AIAA J. 20*, 992-1000, 1982.

[170] Vanderplaats, G.N., *Numerical Optimization Techniques for Engineering Design: With Applications*, McGraw Hill, New York, 1984.

[171] Vanderplaats, G.N., "Thirty Years of Modern Structural Optimization," *Advances in Engineering Software 16,* 81-88, 1993.

[172] Vanderplaats, G.N., and Weisshaar, T.A., "Optimum Design of Composite Structures," *Int. J. Numer. Meth. Engrg 27*, 437-448, 1989.

[173] Vavasis, S.A., "Complexity Issues in Global Optimization: A Survey," in *Handbook of Global Optimization,* edited by R. Horst, and P.M. Pardalos, Kluwer Academic Publishers, Netherlands, 27-41, 1995.

[174] Vinson J.R., and Chou, T-W. *Composite Materials and Their Use in Structures*, John Wiley & Sons, New York, 28-32, 1975.

[175] Visweswaran, V., and Floudas, C.A., "A Global Optimization Algorithm (GOP) for Certain Classes of Nonconvex NLPs - II. Application of Theory and Test Problems," *Computers and Chemical Engineering 14*, 1419, 1990.

[176] Whitney, J.M., *Structural Analysis of Laminated Anisotropic Plates*, Technomic Publ. Co. Inc., Lancaster, PA, 1987.

[177] Wood, G.R., "The Bisection Method in Higher Dimensions," *Mathematical Programming 55 (3)*, 319-338, 1992.

[178] Wood, G.R., Alexander, D.L.J., and Bulger, D.W., "Approximation of the Distribution of Convergence Times for Stochastic Global Optimisation," *Journal of Global Optimization 22*, 271-284, 2002.

[179] Wood, G.R., Bulger, D.W., Baritompa, W.P., and Alexander, D.L.J., "Backtracking Adaptive Search: the Distribution of the Number of Iterations to

Convergence," Technical Report, IIST, Masey University, Palmerston North, New Zealand, 2002.

[180] Wood, G.R., and Zabinsky, Z.B., "Stochastic Adaptive Search," in *Handbook of Global Optimization, Volume 2,* edited by P.M. Pardalos, and H.E. Romeijn, Kluwer Academic Publishers, Netherlands, 231-249, 2002.

[181] Wood, G.R., Zabinsky, Z.B., and Kristinsdottir, B.P., "Hesitant Adaptive Search: the Distribution of the Number of Iterations to Convergence," *Mathematical Programming 89 (3),* 479-486, 2001.

[182] Zabinsky, Z.B., "Global Optimization For Composite Structural Design," *Proceedings of the 35th AIAA/ASME/ASCE/AHS/ASC Structure, Structural Dynamics, and Materials Conference*, April 1994.

[183] Zabinsky, Z.B., "Stochastic Methods for Practical Global Optimization," *Journal of Global Optimization 13,* 433-444, 1999.

[184] Zabinsky, Z.B., "Optimal Design of Composite Structures," *Encyclopedia of Optimization*, Kluwer Academic Publishers, Netherlands, 2002.

[185] Zabinsky, Z.B., Graesser, D.L., Tuttle, M.E., and Kim, G.I., "Global Optimization of Composite Laminates Using Improving Hit-and-Run," in *Recent Advances in Global Optimization*, edited by C.A. Floudas and P.M. Pardalos, Princeton University Press, Princeton, New Jersey, 343-368, 1992.

[186] Zabinsky, Z.B., and Kristinsdottir, B.P., "Complexity Analysis Integrating Pure Adaptive Search (PAS) and Pure Random Search (PRS)," in *Developments in Global Optimization*, edited by Bomze, Csendes, Horst, and Pardalos, Kluwer Academic Publishers, Netherlands, 171-182, 1997.

[187] Zabinsky, Z.B., Seifer, J.D., Tuttle, M.E., Kristinsdottir, B.P., and Neogi, N., "Optimal Design of Composite Panels with Varying Loads," *Proceedings of the ICCE/2 Conference*, New Orleans, 1995.

[188] Zabinsky, Z.B., and Smith, R.L., "Pure Adaptive Search in Global Optimization," *Mathematical Programming 53*, 323-338, 1992.

[189] Zabinsky, Z.B., Smith, R.L., McDonald, J.F., Romeijn, H.E., and Kaufman, D.E., "Improving Hit-and-Run for Global Optimization," *Journal of Global Optimization 3,* 171- 192, 1993.

[190] Zabinsky, Z.B., and Wood, G.R., "Implementation of Stochastic Adaptive Search with Hit-and-Run as a Generator," in *Handbook of Global Optimization, Volume 2,* edited by P.M. Pardalos, and H.E. Romeijn, Kluwer Academic Publishers, Netherlands, 251-273, 2002.

[191] Zabinsky, Z.B., Wood, G.R., Steel, M.A., and Baritompa, W.P., "Pure Adaptive Search for Finite Global Optimization," *Mathematical Programming 69,* 443-448, 1995.

[192] Zangwill, W.I., *Nonlinear Programming: A Unified Approach,* Prentice-Hall, Englewood Cliffs, New Jersey, 1969.

Index

MIX
Papier aus verantwortungsvollen Quellen
Paper from responsible sources
FSC® C105338

If you have any concerns about our products,
you can contact us on
ProductSafety@springernature.com

In case Publisher is established outside the EU,
the EU authorized representative is:
Springer Nature Customer Service Center GmbH
Europaplatz 3, 69115 Heidelberg, Germany

Printed by Libri Plureos GmbH
in Hamburg, Germany